Introduction to Neuroscience

Gary G. Matthews

Department of Neurobiology
State University of New York
Stony Brook, New York

b

**Blackwell
Science**

© 2000 by Blackwell Science, Inc.

Editorial Offices:
Commerce Place, 350 Main Street, Malden, Massachusetts 02148, USA
Osney Mead, Oxford OX2 0EL, England
25 John Street, London WC1N 2BL, England
23 Ainslie Place, Edinburgh EH3 6AJ, Scotland
54 University Street, Carlton, Victoria 3053, Australia
Other Editorial Offices:
Blackwell Wissenschafts-Verlag GmbH, Kurfürstendamm 57, 10707 Berlin, Germany
Blackwell Science KK, MG Kodenmacho Building, 7–10 Kodenmacho Nihombashi, Chuo-ku, Tokyo 104, Japan

Distributors:
USA
    Blackwell Science, Inc.
    Commerce Place
    350 Main Street
    Malden, Massachusetts 02148
    (Telephone orders: 800-215-1000 or 781-388-8250; fax orders: 781-388-8270)

Canada
    Login Brothers Book Company
    324 Saulteaux Crescent
    Winnipeg, Manitoba, R3J 3T2
    (Telephone orders: 204-224-4068)

Australia
    Blackwell Science Pty, Ltd.
    54 University Street
    Carlton, Victoria 3053
    (Telephone orders: 03-9347-0300; fax orders: 03-9349-3016)

Outside North America and Australia
    Blackwell Science, Ltd.
    c/o Marston Book Services, Ltd.
    P.O. Box 269
    Abingdon
    Oxon OX14 4YN
    England
    (Telephone orders: 44-01235-465500; fax orders: 44-01235-465555)

Acquisitions: Nancy Hill-Whilton
Development: Jill Connor
Production: Irene Herlihy
Manufacturing: Lisa Flanagan
Interior design by Colour Mark
Cover design by Madison Design
Typeset by Best-set Typesetter Ltd., Hong Kong
Printed and bound by Capital City Press

Printed in the United States of America
00 01 02 03 5 4 3 2 1

The Blackwell Science logo is a trade mark of Blackwell Science Ltd., registered at the United Kingdom Trade Marks Registry

Library of Congress Cataloging–in–Publication Data

Matthews, Gary G., 1949–
    Introduction to neuroscience / by Gary G. Matthews.
       p.   cm. — (11th hour)
    ISBN 0-632-04414-4
    1. Neurophysiology Examinations, questions, etc.
    2. Neurophysiology Outlines, syllabi, etc. I. Title. II. Series: 11th hour (Malden, Mass.)
    [DNLM: 1. Nervous System. 2. Motor Neurons—physiology.
    3. Neurobiology. 4. Neurons—physiology. 5. Neurons, Afferent—physiology. WL 102 M439i 2000]
    QP356.M34 2000
    612.8′0076—dc21   99-23051
                                                           CIP

Introduction to Neuroscience

11th Hour

CONTENTS

11TH HOUR GUIDE TO SUCCESS

The 11th Hour Series is designed to be used when the textbook doesn't make sense, the course content is tough, or when you just want a better grade in the course. It can be used from the beginning to the end of the course for best results or when cramming for exams. Both professors teaching the course and students who have taken it have reviewed this material to make sure it does what *you* need it to do. The material flows so that the process keeps your mind actively learning. The idea is to cut through the fluff, get to what you need to know, and then help you understand it.

Essential Background. We tell you what information you already need to know to comprehend the topic. You can then review or apply the appropriate concepts to conquer the new material.

Key Points. We highlight the key points of each topic, phrasing them as questions to engage active learning. A brief explanation of the topic follows the points.

Topic Tests. We immediately follow each topic with a brief test so that the topic is reinforced. This helps you prepare for the real thing.

Answers. Answers come right after the tests; but, we take it a step farther (that reinforcement thing again), we explain the answers.

Clinical Correlation or Application. It helps immeasurably to understand academic topics when they are presented in a clinical situation or an everyday, real-world example. We provide one in every chapter.

Demonstration Problem. Some science topics involve a lot of problem solving. Where it's helpful, we demonstrate a typical problem with step-by-step explanation.

Chapter Test. For more reinforcement, there is a test at the end of every chapter that covers all of the topics. The questions are essay, multiple choice, short answer, and true/false to give you plenty of practice and a chance to reinforce the material the way you find easiest. Answers are provided after the test.

Check Your Performance. After the chapter test we provide a performance check to help you spot your weak areas. You will then know if there is something you should look at once more.

Sample Midterms and Final Exams. Practice makes perfect so we give you plenty of opportunity to practice acing those tests.

The Web. Whenever you see this symbol ▣ the author has put something on the Web page that relates to that content. It could be a caution or a hint, an illustration or simply more explanation. You can access the appropriate page through *http://www.blackwellscience.com*. Then click on the title of this book.

The whole flow of this review guide is designed to keep you actively engaged in understanding the material. You'll get what you need fast, and you will reinforce it painlessly. Unfortunately, we can't take the exams for you!

PREFACE

Chapters are divided into two or three topic areas, each starting with a checklist of important questions to be covered and a condensed description of relevant concepts and facts. However, the principal focus of each topic area is the Topic Test, which serves several purposes. First, the Topic Test assesses mastery of the material and pinpoints areas where more work is needed. Second, Topic Test questions sometimes require information not explicitly covered in the text provided here, in order to determine overall knowledge about the topic, rather than immediate recall of the material just read. Third, I have provided not only answers to the questions in the Topic Tests, but also detailed explanations of the answers. These explanations serve a teaching function, of course, but they also allow students to practice making a connection between the wording of a question on an examination and the relevant underlying information required to answer the question.

Each chapter concludes with an overall examination that covers all the material of the chapter. All of the concepts presented in the topic areas and in the Topic Tests of the chapter are included. The Chapter Test is intended to be a second chance for students to test their mastery of the material, after completing the Topic Tests for the chapter. Answers are provided for the questions of the Chapter Test. Each Chapter Test includes at least one essay question, and the essay answer is once again a teaching tool, providing a model of how a professional researcher in neurobiology would answer that question.

Three additional examinations are provided that mimic midterm and final examinations in a neuroscience course. One midterm examination in the middle of the book covers Chapters 1–10, a second midterm examination near the end of the book covers Chapters 11–20, and a final examination at the end of the book covers all 20 chapters. These examinations provide another opportunity for students to find holes in their knowledge and to review material efficiently as they progress through the course.

Students in my own courses frequently ask me where they can find review articles that cover particular aspects of neuroscience. Neuroscience is a rapidly progressing field, and it is important for students to learn how to find the latest reviews on their own to keep up to date. Probably the single best resource for this purpose is the PubMed web site run by the U.S. National Library of Medicine (www.ncbi.nlm.nih.gov/PubMed). The "advanced search" option allows great latitude in finding information and restricting the breadth of the search. Abstracts for most articles are immediately available, and links to full-text versions of the articles are increasingly available, too. I also highly recommend the Annual Review of Neuroscience, a yearly collection of review articles covering a wide range of topics in neurobiology.

I thank the reviewers for their numerous helpful suggestions for improving the presentation and the coverage of this study guide. Neuroscience is a broad discipline, and it is important to get the perspective of others—both instructors and students—regarding the material to be covered. Special thanks go to the student reviewers. After all, this book is intended to help neuroscience students, and their opinions were valuable to me. The reviewers included Wendy I. Bramlett, Sweet Briar College; Daniel M. Chandler, Amherst College; Patsy Dickinson, Bowdoin; Stephen George, Amherst College; Suzanne Giordano, Cedar Crest College; W. R. Klemm, Texas A&M University; Meghna Misra,

Amherst College; James A. Murray, Colby College; Stefan Pulver, Colby College; Janice Naegele, Wesleyan University; and David Stoney, The Medical College of Georgia.

UNIT I:

ORGANIZATION OF THE NERVOUS SYSTEM

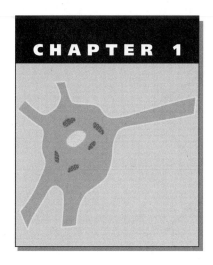

CHAPTER 1

The Vertebrate Nervous System

The nervous system is divided into the **central nervous system** (CNS) and the **peripheral nervous system** (PNS) (**Figure 1.1**). The PNS consists largely of the nerves that carry sensory information to the CNS and motor commands from the CNS. The CNS consists of the **brain** and **spinal cord**, which contain most nerve cells (neurons) of the nervous system, although accumulations of neurons, called **ganglia** (singular = **ganglion**), are found in the PNS as well. Neurons that carry sensory information about the environment or the body are called **sensory neurons**, and neurons that carry signals out of the nervous system and contact targets such as muscles are called **motor neurons**.

ESSENTIAL BACKGROUND

- **General anatomy of vertebrates**
- **Vertebral column and vertebral segments**

TOPIC 1: THE PERIPHERAL NERVOUS SYSTEM

KEY POINTS

✓ *What are the two divisions of the peripheral nervous system?*

✓ *Where are the motor neurons located for the two divisions?*

✓ *What neurotransmitter substances are used by motor neurons?*

The PNS is divided into two parts: the **somatic nervous system** and the **autonomic nervous system**. Somatic motor neurons contact skeletal muscle cells, whereas autonomic motor neurons target gland cells, cardiac muscle cells, and smooth muscle cells. In the somatic nervous system, the cell bodies of motor neurons are located in the CNS, either in the spinal cord or the brain stem. The cell bodies of motor neurons in the autonomic nervous system are located outside the CNS, in autonomic ganglia distributed throughout the body. The autonomic ganglia are controlled by **preganglionic neurons**, located in the spinal cord and brain stem.

The autonomic nervous system is divided into two parts: the **sympathetic** and **parasympathetic** divisions. Sympathetic motor neurons are located in two chains of **paravertebral ganglia** (the **sympathetic chains**) that run parallel to the spinal column, with one pair of ganglia for each vertebral segment, or in **prevertebral ganglia** located in the abdomen. The **parasympathetic ganglia** are distributed more diffusely throughout the body and tend to be located relatively close to their target organs—in some cases, within the target organ itself.

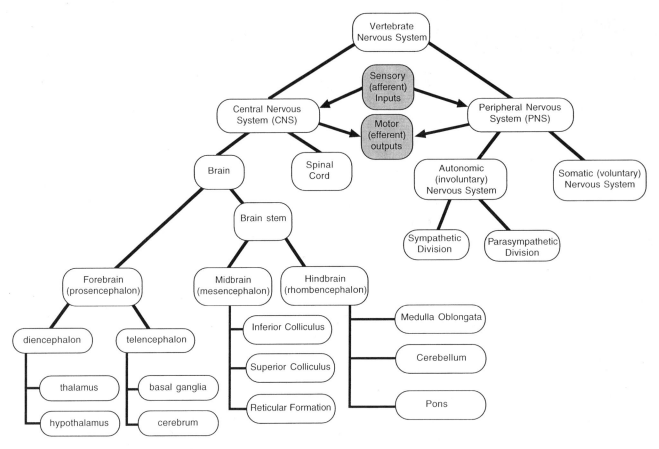

Figure 1.1 The organization of the vertebrate nervous system.

For the most part, the motor neurons of the two autonomic divisions have opposing effects on target organs. For example, the parasympathetic motor neurons slow the heart rate, but the sympathetic motor neurons speed the heart rate. The sympathetic and parasympathetic motor neurons exert opposing actions on target cells by releasing two different chemical neurotransmitter substances, **norepinephrine** for the sympathetic division and **acetylcholine** for the parasympathetic division. All somatic motor neurons use the neurotransmitter acetylcholine, which excites skeletal muscle cells.

Topic Test 1: The Peripheral Nervous System

True/False

1. The two subdivisions of the peripheral nervous system are the brain and spinal cord.

2. Motor neurons carry signals out of the nervous system and make contact with target cells controlled by the nervous system.

Multiple Choice

3. Which of the following is not part of the autonomic nervous system?
   a. parasympathetic ganglia
   b. paravertebral sympathetic ganglia

c. prevertebral sympathetic ganglia

d. somatic motor neurons

4. All motor neurons in both the somatic and autonomic nervous systems
    a. use the neurotransmitter acetylcholine exclusively.
    b. have cell bodies located only in peripheral ganglia situated throughout the body.
    c. may either excite or inhibit their targets.
    d. make contact only with skeletal muscle cells and do not target other cell types.
    e. none of the above

Short Answer

5. List at least three ways in which the somatic and autonomic nervous systems differ.

Topic Test 1: Answers

1. **False.** The two subdivisions of the PNS are the somatic and the autonomic nervous systems. The brain and spinal cord are subdivisions of the CNS.

2. **True.** Motor neurons provide the outputs from the nervous system and are responsible for controlling the various organs and tissues regulated by the nervous system.

3. **d.** Somatic motor neurons are part of the somatic nervous system, not the autonomic nervous system. The paravertebral and prevertebral sympathetic ganglia are the two types of ganglia found in the sympathetic division of the autonomic nervous system. The parasympathetic ganglia are part of the other major division of the autonomic nervous system, the parasympathetic division.

4. **e.** Answers a through d are all false. a. Although all somatic motor neurons in vertebrates use the neurotransmitter acetylcholine, autonomic motor neurons may use either nor-epinephrine or acetylcholine. b. Somatic motor neurons have cell bodies in the spinal cord or brain. c. Somatic motor neurons always excite their targets, the skeletal muscle cells. d. Somatic motor neurons contact skeletal muscle cells, but autonomic motor neurons contact many targets, including endocrine cells and smooth muscle cells.

5. 1) Autonomic motor neurons are located in ganglia outside the CNS, but the cell bodies of somatic motor neurons are located within the CNS. 2) Motor neurons of the autonomic nervous system may either excite or inhibit their targets, but motor neurons of the somatic nervous system always excite their target. 3) Motor neurons of the somatic nervous system control skeletal muscles, but motor neurons of the autonomic nervous system control a variety of targets. 4) Motor neurons of the somatic nervous system release acetylcholine, which excites skeletal muscle cells. Motor neurons of the autonomic nervous system release either acetylcholine or norepinephrine, which typically have opposing actions on target cells.

TOPIC 2: THE CENTRAL NERVOUS SYSTEM

KEY POINTS

✓ *What are the main divisions of the vertebrate CNS?*

✓ *What are the principal subdivisions of the brain?*

✓ *What are the ventricles of the CNS and how are they organized?*

The simplest nervous systems are found in organisms such as *Hydra*, in which a diffuse nerve net spreads throughout the animal's body. During the course of evolution, a major trend is toward increasing centralization of the nervous system, with the neurons collecting into localized ganglia. The ganglia in turn coalesce to form a true CNS. For example, in more complex animals such as annelids and insects, a central nerve cord connects a series of ganglia arrayed longitudinally along the midline.

As more complex nervous systems evolved, the ganglia at the head of the animal became more prominent and fused to become the brain. This trend toward increased size and importance of the ganglia at the head is called **cephalization**. Although it may not be readily apparent, many different clusters of neurons (called **nuclei** or **ganglia**) exist within the mammalian brain, which represents a more intricate arrangement of the clearly separate ganglia found in simpler animals.

The vertebrate CNS is divided into the brain and the spinal cord. The brain consists of three main regions: the **hindbrain** (or **rhombencephalon**), the **midbrain** (or **mesencephalon**), and the **forebrain** (or **prosencephalon**). Collectively, the midbrain and hindbrain form the **brain stem**. The hindbrain is the most posterior part of the brain (the posterior direction—toward the tail—is also called **caudal**) and has three principal divisions: the **medulla oblongata**, **cerebellum**, and **pons**. The medulla oblongata sits at the junction between the brain and the spinal cord and contains neurons controlling visceral functions, such as breathing, swallowing, and digestion. The pons is located on the bottom surface (**ventral** surface) of the brain at the junction between the hindbrain and the midbrain and also regulates respiration. The cerebellum balloons out from the top surface (**dorsal** surface) of the hindbrain and helps coordinate motor commands.

In the midbrain, two important parts are the **inferior colliculus**, which is part of the auditory system, and the **superior colliculus**, which processes visual information. Other parts of the midbrain form part of the **reticular formation**, which controls arousal and plays roles in various sensory and motor systems.

The forebrain is the most anterior part of the brain, toward the front end of the animal (the **rostral** direction). The forebrain is divided into the **diencephalon** and the **telencephalon**. The **thalamus** and **hypothalamus** make up the diencephalon, and the telencephalon consists of **cerebrum** and the **basal ganglia**. The thalamus is primarily a sensory structure that distributes sensory information to the appropriate parts of the cortex. The hypothalamus has a number of subdivisions concerned with homeostasis (hunger, thirst, and thermoregulation), sexual behavior, and emotion. The cerebrum is a major part of mammalian brains, and the **cerebral cortex** dominates the exterior view of the mammalian brain. In mammals more complex than rodents, the surface area of the cerebral cortex will not fit within the skull without forming pronounced infoldings, called **convolutions**. The cerebral cortex is involved in the formation of sensory perceptions, planning and initiation of movement, and associative learning. The other part of the telencephalon, the basal ganglia, helps to control movement.

During development, the CNS forms from a fluid-filled tube (see Chapter 2). In the adult brain, the fluid-filled core remains, forming the **spinal canal** in the spinal cord and the **cerebral ventricles** in the brain. The fluid filling the core is the **cerebrospinal fluid**. As it enters the hindbrain, spinal canal expands to form the **fourth ventricle**. Through the midbrain the ventricle narrows again to form the **cerebral aqueduct**. In the diencephalon, the aqueduct opens dorsally and ventrally to form the **third ventricle**. In the telencephalon, the ventricles expand laterally to form the two **lateral ventricles**, one in each hemisphere.

Topic Test 2: The Central Nervous System

True/False

1. Cephalization refers to the evolutionary trend toward larger nerve nets in animals with larger bodies.

2. The two main divisions of the CNS are the brain and the hindbrain.

3. The three principal divisions of the brain are the rhombencephalon, the mesencephalon, and the prosencephalon.

4. The brain stem is a collective term for the hindbrain and the midbrain.

Multiple Choice

5. The three subdivisions of the hindbrain are
   a. diencephalon, telencephalon, and cerebrum.
   b. medulla oblongata, cerebellum, and pons.
   c. inferior colliculus, superior colliculus, and reticular formation.
   d. thalamus, hypothalamus, and basal ganglia.
   e. none of the above

6. Which of the following is not part of the forebrain?
   a. cerebral cortex
   b. diencephalon
   c. cerebellum
   d. basal ganglia
   e. cerebrum

Short Answer

7. List the important divisions and subdivisions of the forebrain.

8. What are the subdivisions of the brain stem?

Topic Test 2: Answers

1. **False.** Cephalization is the evolutionary trend toward increasing complexity of the neural ganglia at the head of the animal. Nerve nets are a primitive form of nervous system found only in simple organisms such as Hydra.

2. **False.** The two main divisions of the CNS are the brain and the spinal cord.

3. **True.** Other names for the three parts are the hindbrain (rhombencephalon), midbrain (mesencephalon), and forebrain (telencephalon).

4. **True.**

5. **b.** The diencephalon, telencephalon, and cerebrum (choice a) and the thalamus, hypothalamus, and basal ganglia (choice d) are all part of the forebrain, whereas the inferior colliculus, superior colliculus, and reticular formation are associated with the midbrain (choice c).

6. **c.** The cerebellum is part of the hindbrain. All other structures are divisions or subdivisions of the forebrain.

7. The two major divisions of the forebrain are the diencephalon and the telencephalon. The diencephalon is divided into the thalamus and hypothalamus, and the telencephalon is divided into the cerebrum and the basal ganglia. The outer surface of the cerebrum is the cerebral cortex.

8. The brain stem consists of the hindbrain and the midbrain. The hindbrain is subdivided into the medulla oblongata, pons, and cerebellum. The midbrain contains the inferior and superior colliculi (plural of colliculus = colliculi) and parts of the reticular formation.

IN THE CLINIC

Because the nervous system plays such an important role in the body, damage to the nervous system as a result of disease or accident produces clinically significant symptoms. The nature of the symptoms can give valuable clues about where the damage has occurred. In fact, before the advent of noninvasive imaging techniques, such as computed tomography and magnetic resonance imaging, neurological symptoms were the primary means of determining the locus of damage. Also, knowledge of the functional organization of the nervous system guides where a surgeon should or should not cut during neurosurgery. A simple example is the loss of sensation and motor control when a peripheral nerve is cut: only the part of the body that is innervated by the damaged nerve is affected. Mapping between particular nerves and particular body regions is well established, and so the locus of the affected region establishes which nerve branch is damaged. In the CNS, the situation is often less simple, because of the anatomical complexity and because damage produced by a stroke or a tumor typically does not respect anatomical boundaries between brain regions. Nevertheless, neurologists are able to distinguish among different syndromes that result from neurological problems. For example, damage to the cerebellum might produce disorders of balance or might disrupt fine motor movements, depending on the part of the cerebellum that is damaged. In the cerebral cortex, certain regions process specific types of sensory information, and lesions in those regions disrupt perception of the corresponding sensory modality. In the brain stem, damage is often fatal, because of the importance of neural circuits in the brain stem to vital functions like respiration.

Chapter Test

True/False

1. Motor neurons carry commands (motor commands) from the nervous system to the target tissues controlled by the nervous system.

2. The vertebrate CNS consists of the brain and spinal cord.

3. The autonomic nervous system consists of the somatic subdivision and the sympathetic subdivision.

4. All somatic motor neurons release the neurotransmitter norepinephrine.

5. In the autonomic nervous system, some motor neurons release the transmitter acetylcholine, but other motor neurons release norepinephrine.

6. The hindbrain, midbrain, and forebrain are the three major divisions of the vertebrate brain.

7. The brain stem is a subdivision of the hindbrain.

8. The cerebellum is part of the cerebral cortex.

9. The basal ganglia are part of the PNS.

10. The diencephalon and the telencephalon are the two major divisions of the forebrain.

Multiple Choice

11. Which of the following statements is not true of the cerebral ventricles? The cerebral ventricles
    a. are fluid-filled passages at the core of the brain.
    b. are filled with cerebrospinal fluid.
    c. are found only in the forebrain.
    d. are divided into the lateral ventricles, third ventricle, cerebral aqueduct, and fourth ventricle.

12. Which of the following statements is not true of the PNS?
    a. The PNS is divided into the autonomic and somatic nervous systems.
    b. The PNS consists solely of sensory neurons and contains no motor neurons.
    c. The PNS includes the sympathetic and the parasympathetic ganglia.
    d. The PNS includes the nerves that connect the body with the CNS.

13. The autonomic nervous system
    a. consists of sympathetic and parasympathetic divisions.
    b. provides involuntary control of a variety of targets, including endocrine glands, smooth muscle, and cardiac muscle.
    c. includes the paravertebral sympathetic ganglia, prevertebral sympathetic ganglia, and parasympathetic ganglia.
    d. has motor neurons that release either norepinephrine (sympathetic) or acetylcholine (parasympathetic).
    e. all of the above are correct.

14. The CNS
    a. consists of the brain and the autonomic ganglia.
    b. contains the cell bodies of somatic motor neurons.
    c. does not include the spinal cord.
    d. consists of the forebrain but not the midbrain and hindbrain.
    e. none of the above is correct.

15. Select the correct statement:
    a. The diencephalon and the telencephalon are subdivisions of the midbrain.
    b. The cerebellum and the pons are part of the forebrain.
    c. The medulla oblongata is part of the brain stem.
    d. The cerebral cortex is part of the PNS.
    e. All of the above are correct.

Short Answer/Short Essay

16. The _____ division of the autonomic nervous system uses the neurotransmitter norepinephrine and the _____ division of the autonomic nervous system uses the neurotransmitter acetylcholine.

17. Name the parts of the fluid-filled core of the CNS in sequence, beginning with the spinal cord.

18. Give brief definitions of the following:
    a. ganglia
    b. brain stem
    c. cephalization
    d. prosencephalon

19. Briefly describe some of the functions of the three subdivisions of the hindbrain.

20. What are the parts of the forebrain and what are their functions?

Chapter Test Answers

1. **T** 2. **T** 3. **F** 4. **F** 5. **T** 6. **T** 7. **F** 8. **F** 9. **F** 10. **T** 11. **c** 12. **b**
13. **e** 14. **b** 15. **c**

16. sympathetic, parasympathetic

17. In the spinal cord, the fluid-filled core is the spinal canal. In the hindbrain, the canal expands to become the fourth ventricle, which then narrows again to form the cerebral aqueduct in the midbrain. In the diencephalon is the third ventricle, and the lateral ventricles are found in the cerebral hemispheres.

18. **a.** Ganglia are groups or clusters of nerve cells, either in the PNS (e.g., sympathetic ganglia) or in the CNS (e.g., basal ganglia).

    b. The brain stem is a collective term for the midbrain and hindbrain of the vertebrate brain.

    c. Cephalization is the evolutionary trend toward increasing complexity and importance of the ganglia in the head of an organism.

    d. Prosencephalon is another name for the forebrain, which consists of the diencephalon (thalamus and hypothalamus) and telencephalon (basal ganglia and cerebrum).

19. The medulla oblongata controls many visceral functions, such as breathing, swallowing, and digestion. The pons also includes neurons that help to regulate respiration. The cerebellum is a motor coordination center.

20. The two main divisions of the forebrain (prosencephalon) are the diencephalon and the telencephalon. The diencephalon in turn consists of the thalamus (a sensory relay center) and the hypothalamus, which is concerned with homeostasis, emotion, and sexual behavior. The telencephalon consists of the basal ganglia, which are involved in the control of movement, and the cerebrum. The cerebrum mediates higher functions, including sensory perception, planning and initiation of movement, and associative learning.

Check Your Performance:

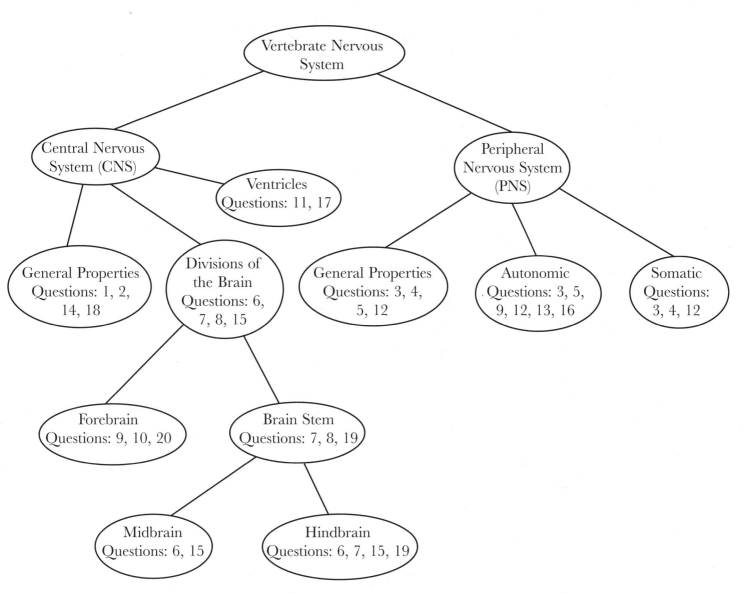

Note the number of questions in each grouping that you got wrong on the chapter test. Identify areas where you need further review and go back to relevant parts of this chapter.

Further help: If you continue to have difficulty with the concepts, a detailed presentation of these topics can be found in chapter 2 of *Neurobiology: Molecules, Cells, and Systems*, published by Blackwell Science, Inc.

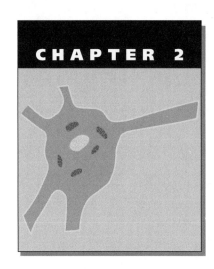

CHAPTER 2

Development of the Vertebrate Nervous System

The divisions among the various regions of the brain may seem rather arbitrary when we view the overall structure of the adult brain. Why is the brain divided into three major divisions (forebrain, midbrain, and hindbrain) instead of some other number? How are the dividing lines between the three divisions set? Why is the forebrain further subdivided into the telencephalon and the diencephalon? Examining the embryogenesis of the brain gives some answers to these questions.

ESSENTIAL BACKGROUND

- **General anatomy of vertebrates**
- **Basic embryology of vertebrates**
- **Cell proliferation and the cell cycle**

TOPIC 1: NEURAL INDUCTION IN THE EARLY EMBRYO

KEY POINTS

✓ *How does the nervous system arise during gastrulation?*

✓ *What mechanisms might be involved in neural induction?*

✓ *What are some signal proteins that regulate neural induction?*

As the fertilized egg divides, different progeny cells give rise to internal organs, muscles, and skin. Among the latter group, a subpopulation gives rise to neurons and glial cells of the nervous system. Vertebrate embryos have three layers of cells: an inner layer (**endoderm**), a middle layer (**mesoderm**), and an outer layer (**ectoderm**). These layers form from the single-layer early embryo by a complex infolding process, called **gastrulation**.

The nervous system arises from part of the ectoderm called the **neuroectoderm**, and the selection of part of the ectoderm to become neural tissue is termed **neural induction**. During gastrulation, the mesoderm forms the **notochord**, which is a rod-shaped structure along the midline of the developing embryo, defining the longitudinal axis of the body. Neuroectoderm is induced in the ectoderm lying directly over the developing notochord. The part of the embryo that generates the notochord is called the **Spemann organizer**.

Induction of neuroectoderm requires diffusible messenger molecules, whose identity is not yet firmly established. One possibility is that the notochord releases an **inducer substance**, which stimulates formation of neural tissue by cells that would otherwise become other tissues (e.g., skin). One candidate for an inducer substance is the protein **noggin**, which induces excessive brain and head development when injected into frog embryos. Most likely, noggin is only one of several soluble inducing factors released at various times during development of the nervous system, but it appears to be involved in early stages of neural induction. Another possibility is that the notochord releases a **suppressor substance**, which interferes with signals that cause ectoderm to form non-neural tissue. In a suppressor mechanism, neural tissue is considered the default state for ectoderm.

An example of a suppressor mechanism is the interaction between the signal proteins **activin** and **follistatin**. Activin is released from endodermal cells and induces formation of mesoderm by ectoderm, which would otherwise form neural tissue. However, cells of the notochord release follistatin, which inactivates activin. So, activin levels are reduced in the vicinity of the notochord, and loss of activin causes ectoderm cells near the notochord to form neural tissue (the default state). Both suppressor and inducer mechanisms may be involved in neural development. Suppression of activin may be required together with release of an inducer (e.g., noggin) to cause proper formation of neural structures by the embryo.

Topic Test 1: Neural Induction in the Early Embryo

True/False

1. The nervous system develops from part of the ectoderm called the neuroectoderm.

2. The neuroectoderm induces formation of the notochord in underlying mesoderm.

Multiple Choice

3. Induction of neuroectoderm
   a. is mediated by a diffusible signal generated by the Spemann organizer.
   b. may involve both suppressor substances and inducer substances.
   c. is carried out by part of the mesoderm called the notochord.
   d. occurs during gastrulation.
   e. all of the above.

4. According to the suppressor mechanism for neural induction,
   a. development of neural tissue is the default path for ectoderm cells.
   b. release of the inducer protein, noggin, by the endoderm is suppressed by the neuroectoderm.
   c. ectoderm produces skin tissue only in the area overlying the notochord.
   d. the signal protein, activin, is released by the notochord.
   e. all of the above.

Short Answer

5. Suppose we take a piece of notochord from an early embryo and transplant it into another embryo in a position distant from the recipient's own notochord, so that the recipient embryo develops two notochords (its own and the transplanted one). What would happen to the ectoderm overlying the transplanted notochord?

Topic Test 1: Answers

1. **True.** Neural cells arise from the neuroectoderm, which is part of the ectoderm.

2. **False.** In fact, the notochord is responsible for formation of the neuroectoderm, not vice versa.

3. **e.** All of the statements are correct. During gastrulation, as the mesoderm forms, part of the mesoderm along the midline becomes the notochord, which defines the head-tail axis. Substances released by cells of the notochord act on overlying ectodermal cells to elicit development of neural cells. The forming notochord is also called the Spemann organizer, in honor of the biologist who first described the embryonic formation of the nervous system. Neural induction may involve release of substances that induce neural cells and substances that suppress induction of non-neural cells.

4. **a.** According to the suppressor hypothesis, ectoderm cells become neural cells by default and must be stimulated to form non-neural cells by a messenger molecular such as activin. Statement b is incorrect, because although noggin is an inducer protein, it is released by the notochord (part of the mesoderm), not by the endoderm. Also, production of noggin is not inhibited by the neuroectoderm. Statement c is incorrect, because the ectoderm in fact produces neural tissue in the area overlying the notochord. Statement d is incorrect, because activin is produced by the endoderm; the notochord produces the activin inhibitor, follistatin, which reduces activin levels near the notochord.

5. Because neural induction is stimulated by a diffusible signal released by the notochord, the second notochord will induce a second neuroectoderm at the transplant site, where the ectoderm would normally produce skin. The recipient embryo would thus develop two nervous systems: its own and a second at the transplant site. Transplantation experiments of this kind were carried out by Spemann in the early 20th century to establish the importance of the notochord in neural induction and to establish that a diffusible messenger produced by the notochord mediates induction.

TOPIC 2: EARLY DEVELOPMENT OF THE NERVOUS SYSTEM

KEY POINTS

✓ *What are the earliest stages in formation of the vertebrate nervous system?*

✓ *What structures in the early nervous system give rise to the various brain regions?*

After induction of the neuroectoderm, the nervous system forms in a series of stages illustrated in **Figure 2.1**. The neuroectoderm forms two distinct parts: the **neural plate** and the **neural crest** (**Figure 2.1A**). The central nervous system (CNS) arises from the neural plate, and the neural crest produces the peripheral nervous system (PNS). As cells of the neural plate proliferate, they create an indentation called the **neural groove**, with cells of the neural crest forming the top (the crest) of each side of the groove (**Figure 2.1B**). As the indentation deepens, the lips of the groove come closer together and fuse, and the neural groove becomes a sealed tubular structure called the **neural tube**, separate from the overlying ectoderm. Cells of the neural crest lie between the neural tube and the ectoderm (**Figure 2.1C**). The hollow core of the neural tube becomes the spinal canal of the spinal cord and the ventricles of the brain.

Cells of the neural tube differentiate into neurons and glial cells of the brain and spinal cord. At the anterior end of the neural tube, which forms the brain, cells proliferate more rapidly than at

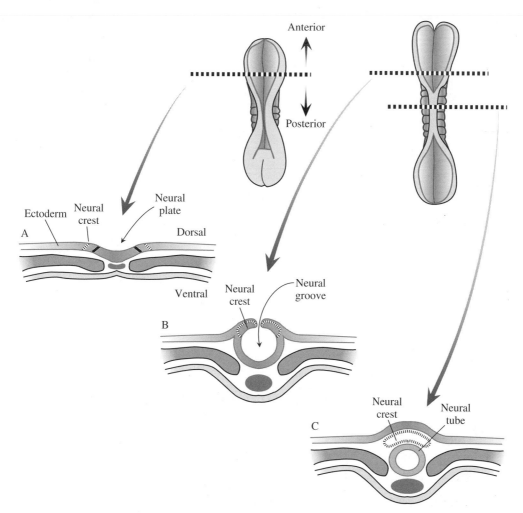

Figure 2.1 Early differentiation of the nervous system (A–C) and development of the brain (D–E) in a vertebrate embryo. A. The nervous system starts out as undifferentiated ectodermal cells along the dorsal midline, called the neural plate. At this stage, the cells that will form the neural crest lie along each side of the neural plate. B. At a later stage, the neural plate invaginates to form the neural groove, with the neural crest along the lips of the groove. C. Still later, the neural groove pinches off to form the neural tube, as the two opposing parts of the neural crest fuse. D. At a primitive stage of development, the anterior neural tube expands into three vesicles: the hindbrain, midbrain, and forebrain. E. As development proceeds, the hindbrain and forebrain vesicles subdivide further.

the posterior end, which forms the spinal cord. Thus, the anterior end bulges out, forming three protrusions called **vesicles**. The three vesicles are called the forebrain, midbrain, and hindbrain (**Figure 2.1D**). The fact that there are three vesicles, separated from each other by constrictions, explains why the adult brain is divided into three major divisions. Assignment of a particular structure in the adult brain to the forebrain, midbrain, or hindbrain is determined by which of the primordial brain vesicles gives rise to the structure. As development proceeds, the lateral portions of the forebrain vesicle balloon out into structures that become the two hemispheres of the cerebrum (**Figure 2.1E**). These dual protrusions form the telencephalon, whereas the remainder of the forebrain vesicle becomes the diencephalon (Figure 2.1E). At this developmental stage, the subdivision of the hindbrain into its components also begins. As the telencephalon grows, it folds back on itself and envelops the midbrain and hindbrain. In animals with large cortical regions, such as primates, the growing telencephalon turns back on itself once again, forming a shape like a ram's horn.

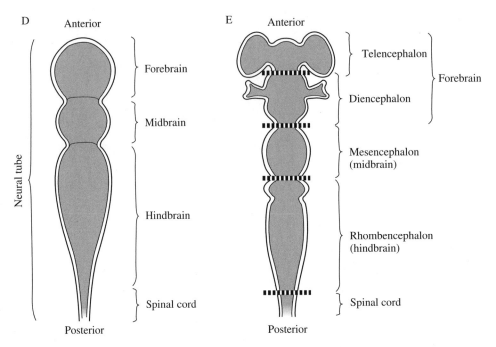

Figure 2.1 *Continued*

Topic Test 2: Early Development of the Nervous System

True/False

1. The neural plate gives rise to the CNS.

2. The PNS arises from the neural crest.

3. The neural crest fuses to become the neural tube.

4. Structures included in the adult forebrain arise from the most anterior vesicle of the neural tube, the forebrain vesicle.

Multiple Choice

5. The three brain vesicles of the neural tube are
   a. diencephalon, telencephalon, and rhombencephalon.
   b. spinal cord, neural crest, and neural plate.
   c. forebrain, midbrain, and hindbrain.
   d. neural groove, neural plate, and neuroectoderm.
   e. none of the above.

6. Which of the following statements is not correct? The neural crest
   a. is part of the neuroectoderm.
   b. produces the PNS.
   c. gives rise to the spinal cord.
   d. is located at the upper lips of the neural groove.
   e. lies between the neural tube and the overlying ectoderm after the neural tube forms.

Short Answer

7. Briefly describe the stages in formation of the neural tube, beginning with induction of neuroectoderm by the notochord.

Topic Test 2: Answers

1. **True.** The neural plate is the central part of the neuroectoderm. As cells of the neural plate proliferate, they pinch off to form the neural tube, which becomes the spinal cord and brain.

2. **True.** The neural crest begins as the lateral portions of the neuroectoderm and is left behind as the neural tube pinches off. The PNS then forms from cells of the neural crest.

3. **False.** The neural tube forms from the neural plate (see Figure 2.1). Neural crest cells are excluded from the neural tube as it pinches off.

4. **True.** At early stages, shortly after formation of the neural tube, the three brain vesicles form at the anterior end of the neural tube. The most anterior vesicle is the forebrain vesicle, from which the structures of the adult forebrain originate.

5. **c.** The three vesicles have the same names as the three major divisions of the adult brain.

6. **c.** The spinal cord is part of the CNS and arises from the neural plate. The neural crest becomes the PNS. As the neural groove forms, the cells of the neural crest are located at the upper lips of the groove (choice d). When the edges of the neural groove fuse, the neural crest separates from the neural tube and lies between the tube and the overlying ectoderm (choice e). Early in development the neural crest forms at the lateral edges of the neuroectoderm (choice a).

7. During gastrulation, cells of the ectoderm adjacent to the developing notochord are induced to form the neuroectoderm. As cells of the neuroectoderm proliferate, they form the neural plate. With time, an indentation forms at the midline, which deepens to become the neural groove. The lips of the groove then fuse and the neural tube pinches off from the remaining ectoderm. At this point, cells of the neural crest separate from the neural tube.

TOPIC 3: NEURONAL PROLIFERATION AND MIGRATION

KEY POINTS

✓ *Where in the developing nervous system are neurons born?*

✓ *How do neurons migrate from their birthplace to the correct position in the nervous system?*

✓ *What mechanisms allow neurons to grow long processes that connect with appropriate target cells?*

Precursor cells of the neural tube give rise to all cells in the CNS. Thus, cells of the neural tube must divide prolifically to supply the large number of cells needed to form the adult brain and spinal cord. Proliferation of cells does not occur uniformly throughout the nervous system. Instead, production of new cells is localized to the **ventricular zone**, which is the innermost part of the neural tube. Early in development, the neural tube consists of only two layers: the ventricular zone, where dividing cells are found, and the **marginal zone**, toward the outer edge of the neural tube.

After cells divide in the ventricular zone, daughter cells take one of two avenues: They re-enter the cell cycle and continue to produce more progeny or they exit from the ventricular zone and migrate to form the nervous system. The departing cells become either neurons, which lose the ability to divide further, or glial precursor cells, which retain the ability to proliferate. As cells migrate to their correct position within the three-dimensional structure of the nervous system, they follow a cellular highway formed by a class of glial cells, the **radial glial cells**, that form early in development. These glial cells have a single long process that extends radially from the ventricular surface to the outermost surface of the neural tube. The migrating neurons entwine around the radial glial process, and this close contact of the two cells is maintained via complementary **cell adhesion molecules** located on the surfaces of the two interacting cells. Different adhesion molecules are expressed by different cells, and cells will adhere to each other only if they have matching adhesion molecules.

After a neuron migrates to its final position within the nervous system, it must create processes (dendrites and axons) that connect to other neurons. Collectively, dendrites and axons are called **neurites**. Neurites often travel long distances to reach their targets. At the leading edge of a growing neurite is a **growth cone**, which is a mobile amoeba-like structure that moves by extending filopodia. Movement of filopodia is driven by the interaction of the proteins **actin** and **myosin**, which also form the contractile apparatus of muscle cells. As the growth cone moves forward, the neurite is laid down along the path traveled by the growth cone. The backbone of the neurite consists of filamentous microtubules, which consist of the protein **tubulin**.

Topic Test 3: Neuronal Proliferation and Migration

True/False

1. Nerve cell progenitors are formed by cell division within the marginal zone of the neural tube.
2. Radial glial cells form the cellular highway along which neural progenitor cells migrate.
3. Growth cones are a type of cell adhesion molecule.

Multiple Choice

4. Migration of neural cells
   a. is necessary because neural cells are born in the neural crest and must migrate into the neural tube.
   b. is guided by the interaction between cell adhesion molecules of the neural cell and the radial glial cells.
   c. is necessary because neurons cannot proliferate until they reach their final position in the brain.
   d. requires the prior formation of neurites, which guide cell migration.
   e. none of the above.

5. Neurites
   a. are the cells that become neurons, rather than glial cells, in the developing brain.
   b. is a collective term for dendrites and axons of neurons.
   c. refer to the radial fibers formed by radial glial cells.
   d. is another term for filopodia of growth cones.
   e. none of the above.

Short Answer

6. Briefly describe the process of neurite outgrowth.

Topic Test 3: Answers

1. **False.** Cell division occurs within the ventricular zone. The marginal zone is a relatively cell-free zone at the outer edge of the neural tube.

2. **True.** Radial glial cells form early in development and have a long fiber extending from the ventricular zone to the outer surface of the neural tube. Neural cells migrate from the ventricular zone along the radial fiber of the glial cells.

3. **False.** Growth cones are amoeboid structures at the end of growing neurites.

4. **b.** Cell division occurs in the ventricular zone (therefore, statements a and c are false), and neurons and glial cells must migrate away from the ventricular zone to take up their final positions in the nervous system. Migration occurs along radial glial cells because complementary cell adhesion molecules on the radial glial cells and migrating cells make the cells stick to each other.

5. **b.** Neurites are the processes (dendrites and axons) of nerve cells.

6. Neurites form along the path traveled by the growth cone, which is located at the leading edge of a growing neurite. Growth cones move by extending filopodia that pull the growth cone forward. The movement of the filopodia is driven by motor proteins actin and myosin. The backbone of the extending neurite behind the growth cone is made up of neurotubules, which consist of the protein tubulin.

IN THE CLINIC

During the early development, the embryo is especially sensitive to both environmental and genetic factors. Because the nervous system begins to form at early stages, malformation of the nervous system is a common outcome of infection, environmental toxins, or genetic abnormality during early embryogenesis. Severity of the effect depends on when during development the problem arises and on which part of the developing nervous system is affected. Particularly severe defects arise from failure of the neural tube to close properly (neural tube defects). If the anterior neural tube fails to fuse, the vesicles that give rise to the brain are malformed, producing **anencephaly** (absence of the brain). If more posterior parts of the neural tube fail to fuse properly, the resulting defect is called **spina bifida**. Neuronal communication within the spinal cord is disrupted posterior to the site of the defect, producing varying degrees of paralysis. Defects can also arise later in development, after formation of the neural tube. For example, the forebrain vesicle may fail to form two telencephalic protrusions, resulting in a single telencephalon (**holoprosencephaly**). Severe mental retardation is the outcome.

Chapter Test

True/False

1. The neuroectoderm is the part of the ectoderm that gives rise to the nervous system.

2. Induction of the neuroectoderm occurs before gastrulation, during the stage when the embryo consists of a single cellular layer.

3. Noggin, activin, and follistatin are signaling proteins involved in neural induction.

4. The Spemann organizer is the part of the mesoderm that will form the notochord, which is the structure that induces neural tissue in the overlying ectoderm.

5. The neural tube and the neural groove are two functional divisions of the neuroectoderm.

6. The PNS arises from the neural crest.

7. The brain develops from three vesicles at the anterior end of the neural tube.

8. In the neural tube, neural cells divide within the marginal zone and migrate into the ventricular zone.

9. Radial glial cells are the last cells to develop in the nervous system.

10. Growth cones form at the tips of neurites and are responsible for moving the growing neurite forward.

Multiple Choice

11. Neural induction
    a. occurs simultaneously in all parts of the ectoderm.
    b. is triggered by the signal protein activin, which causes ectodermal cells to form neural cells.
    c. involves the formation of the neural tube by the mesoderm.
    d. is the default differentiation pathway of the endoderm.
    e. may involve both the induction of neural cells and the suppression of non-neural cells in the neuroectoderm.

12. The neural tube
    a. forms from the fusion of the upper lips of the neural groove.
    b. gives rise to the brain and spinal cord.
    c. produces three vesicles at its anterior end, which become the forebrain, midbrain, and hindbrain.
    d. all of the above.

13. In the neural tube, neurons are produced by cell division
    a. throughout all parts of the neural tube.
    b. in the marginal zone of the neural tube.
    c. in the ventricular zone of the neural tube.
    d. of the radial glial cells.

14. Growth cones
    a. move along the fibers of radial glial cells.
    b. are found at the ends of growing neurites.
    c. are characteristic of glial cells but not of neurons.
    d. shuttle migrating neurons to and from the ventricular zone of the neural tube.

Short Answer

15. The _____ gives rise to the cells of the PNS, whereas the _____ gives rise to the cells of the CNS.

16. Identify the following:
    a. tubulin
    b. follistatin
    c. noggin

17. Briefly describe the following:
    a. cell adhesion molecule
    b. radial glial cell
    c. ventricular zone
    d. filopodia

18. Define the following:
    a. neural induction
    b. notochord
    c. Spemann organizer

Short Essay

19. Concisely describe the sequence of events that lead to the formation of the forebrain, starting with gastrulation.
20. Compare and contrast the inducer and suppressor mechanisms proposed for neural induction during embryogenesis. Provide specific examples of signaling proteins that might be involved in these mechanisms.

Chapter Test Answers

1. **T** 2. **F** 3. **T** 4. **T** 5. **F** 6. **T** 7. **T** 8. **F** 9. **F** 10. **T** 11. **e** 12. **d**

13. **c** 14. **b**

15. neural crest, neural plate (neural tube would also be acceptable for the latter)

16. a. Tubulin is the protein that makes up neurotubules, which form the backbone of neurites. b. Follistatin is a protein secreted by the notochord that inactivates the signal protein activin. c. Noggin is a possible inducing factor for neural induction, secreted by the notochord.

17. a. Cell adhesion molecules are molecules on the surface of cells that can interact with complementary adhesion molecules on other cells, allowing the cells to maintain contact with each other. b. Radial glial cells arise early in neural tube development and extend a long fiber from the ventricular zone at the center of the neural tube to the outer margin of the neural tube. Migrating nerve cells follow these radial fibers away from the ventricular zone. c. The ventricular zone is the inner most part of the neural tube. Cell division to produce neurons and glial cells occurs in the ventricular zone. d. Filopodia are the finger-like extensions of growth cones that propel the forward motion of the growth cone during neurite growth.

18. a. Neural induction refers to the process by which ectodermal cells along the midline of the developing embryo are induced to form neural cells rather than other tissue. b. The notochord is the portion of the developing mesoderm that lies along the longitudinal midline of the embryo and induces the formation of neural tissue in the overlying ectoderm. c. The Spemann organizer is part of the mesoderm that will form the notochord. The organizer induces the formation of neural tissue.

19. During gastrulation, the mesoderm forms as an infolding of the embryo. As the mesoderm forms, the cells along the longitudinal midline of the embryo become the notochord, which releases substances that induce nearby cells of the ectoderm to form neural tissue (the neuroectoderm). The central part of the neuroectoderm is the neural plate, which becomes the CNS. As the cells of the neural plate proliferate, they form a groove along the midline (the neural groove), and the upper lips of the groove merge and separate the neural tube from the surrounding ectoderm. The cells at the anterior end of the neural tube proliferate more rapidly and form three protrusions, called vesicles. The most anterior vesicle is the forebrain vesicle, which becomes the forebrain. The

telencephalon develops from the front part of the forebrain vesicle, which forms two protrusions that become the paired cerebral hemispheres of the forebrain.

20. During neural induction, the notochord releases a soluble message that diffuses to the nearby parts of the ectoderm and causes the ectodermal cells to differentiate into neural cells. The suppressor and inducer hypotheses differ in how the ectodermal cells are thought to be affected by the soluble message. In the suppressor hypothesis, the ectodermal cells are assumed to be destined to become neural cells, unless they receive a signal that stimulates them to become non-neural cells. In the inducer hypothesis, it is assumed that ectodermal cells must be actively stimulated by some signal to become neural cells. An example of a possible inducer substance is noggin, which is a protein molecule made by cells of the notochord and secreted into the extracellular space. Noggin stimulates excessive production of neural tissue by the ectoderm when it is injected into embryos. An example of a possible suppressor of neural differentiation by the ectoderm is activin, which is released by the endoderm and stimulates formation of mesoderm by the ectoderm. The notochord is thought to release a substance that locally inactivates activin, preventing it from stimulating the nearby ectoderm to become non-neural cells. The inactivator of activin may be follistatin, which binds to activin and prevents it from reaching the ectoderm in the vicinity of the notochord. It is likely that both active induction of neural differentiation (e.g., molecules like noggin) and suppression of influences that promote non-neural differentiation (e.g., the follistatin/activin interaction) are involved in neural induction.

Check Your Performance:

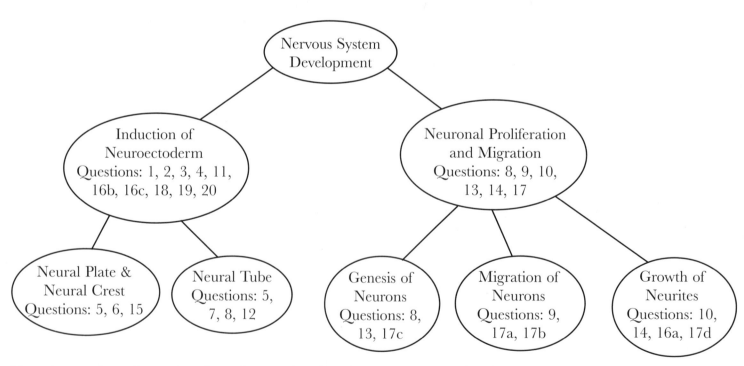

Note the number of questions in each grouping that you got wrong on the chapter test. Identify areas where you need further review and go back to relevant parts of this chapter.

Further help: If you continue to have difficulty with the concepts, a detailed presentation of these topics can be found in chapters 2 and 20 of *Neurobiology: Molecules, Cells, and Systems*, published by Blackwell Science, Inc.

UNIT II:
ELECTRICAL PROPERTIES OF NERVE CELLS

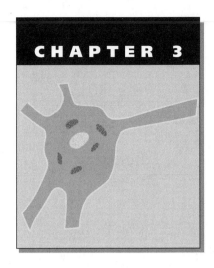

CHAPTER 3

Origin of Electrical Membrane Potential

All cells have a steady electrical voltage across the plasma membrane, called the **resting membrane potential**, which is typically about −70 mV in a neuron (mV = millivolt; 1 volt = 1,000 mV). The minus sign indicates that the inside of the cell is more negative than the outside. The membrane potential arises from the distribution of charged substances (**ions**) in the intracellular and extracellular fluids and from the permeability of the plasma membrane to the different ions. The **extracellular fluid (ECF)** has high concentrations of positively charged sodium ions (positively charged ion = **cation**) and negatively charged chloride ions (negatively charged ion = **anion**) but a low concentration of potassium cations. In contrast, the **intracellular fluid (ICF)** has low sodium and chloride but high potassium and other organic and inorganic anions (proteins, amino acids, sulfate, phosphate, etc.). We first review the principles that govern thermodynamic equilibrium for ions across the membrane. Then we review nonequilibrium steady-state conditions that apply to the resting membrane potential.

ESSENTIAL BACKGROUND

- Osmolarity of solutions
- Diffusion of substances in solution
- Dissociation of salts in solution
- Logarithms (base 10 and base e)
- ATP as a cellular energy source and ATPases as enzymes that release that energy
- Structure and composition of the plasma membrane
- Electrical voltage and electrical current

TOPIC 1: IONIC EQUILIBRIUM

KEY POINTS

✓ *What role does selective permeability of the plasma membrane play in generation of the membrane potential?*

✓ *What conditions govern equilibrium of a charged substance across the plasma membrane?*

✓ *What equation is used to calculate the equilibrated membrane potential for a permeant ion?*

Typical ionic compositions of the ECF and ICF for an animal cell are shown in **Figure 3.1**. Consider the diffusion of potassium ions (K⁺). Most cell membranes have high permeability to

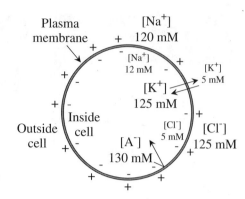

Figure 3.1 Diagram of the ionic composition of the intracellular and extracellular fluid for a mammalian cell.

K^+, and a large concentration gradient drives movement of K^+ out of the cell. As K^+ leaves the cell, positive charge is lost from the cell. Internal anions (A^-) are predominantly organic substances (proteins and amino acids) that cannot cross the plasma membrane. Because the negative charge is trapped within the cell, efflux of positive K^+ is not balanced by efflux of negative charge, and the interior of the cell becomes negative. This electrical gradient also influences the K^+ movement across the membrane, attracting K^+ to the cell interior. Thus, two competing fluxes of K^+ arise (**Figure 3.1**): efflux down the concentration gradient and influx down the electrical gradient. The negativity inside the cell will reach a steady level when these two competing fluxes are exactly equal and opposite. At that point, the electrical gradient exactly balances the concentration gradient, and the system is in equilibrium.

The voltage across the membrane at equilibrium is called the **equilibrium potential**. The equation used to calculate the equilibrium potential is the **Nernst equation**:

$$E_K = \frac{RT}{ZF}\ln\left(\frac{[K^+]_o}{[K^+]_i}\right)$$
(3.1)

In this equation, E_K is the voltage at equilibrium (the potassium equilibrium potential), R is the gas constant, T is the absolute temperature, Z is the valence of the ion (+1 for potassium), F is Faraday's constant, "ln" is the symbol for natural (base e logarithm), and $[K^+]_o$ and $[K^+]_i$ are the K^+ concentrations outside and inside the cell.

A simplified form of the Nernst equation is commonly used:

$$E_K = \frac{58\,mV}{Z}\log\left(\frac{[K^+]_o}{[K^+]_i}\right)$$
(3.2)

The constant 58 mV in this equation arises from converting from natural to base 10 logarithms, evaluating RT/F at room temperature (20°C) and expressing the result in millivolts. From Eq. (3.2) we calculate that the K^+ equilibrium potential in Figure 3.1 is about −81 mV. If K^+ were the only permeant ion, the membrane potential would reach this equilibrated value and then remain there, without any expenditure of energy.

Topic Test 1: Ionic Equilibrium

True/False

1. Movement of an ion across the cell membrane is determined by two factors: the concentration gradient for the ion and the voltage across the membrane.

2. At the equilibrium potential for a permeant ion, flux of the ion down its concentration gradient is greater than flux of the ion down the electrical gradient.

3. The equilibrium potential for a permeant ion is calculated from the Nernst equation, which gives the membrane potential at which there is no net flux of the ion across the membrane.

Multiple Choice

4. Suppose a cell is permeable to K^+, with internal $K^+ = 100\,mM$ and external $K^+ = 10\,mM$. If the membrane potential of the cell were $-58\,mV$, K^+ would move in which direction across the membrane?
   a. into the cell
   b. out of the cell
   c. no net movement

5. Suppose the equilibrium potential for a cation is positive (i.e., the inside of the cell is positive with respect to the outside at equilibrium). This requires that the internal concentration of the cation must be
   a. greater than the external concentration.
   b. less than the external concentration.
   c. equal to the external concentration.

Short Answer

6. If the cell shown in Figure 3.1 were permeable to sodium ions (Na^+), write the Nernst equation for Na^+ and calculate the Na^+ equilibrium potential.

7. A cell is permeable to K^+, and the external and internal K^+ concentrations are equal. What is the K^+ equilibrium potential?

Topic Test 1: Answers

1. **True.** Dissolved substances in solution move down concentration gradients. A charged substance is also sensitive to voltage gradients. Thus, movement of an ion across a cell membrane depends on both the concentration difference and the voltage difference across the membrane.

2. **False.** By definition, the equilibrium potential is the membrane potential at which movement of a permeant ion down its concentration gradient is exactly balanced by movement down the voltage gradient. Thus, at the equilibrium potential, the net movement of the ion must be zero.

3. **True.** The Nernst equation provides the quantitative balance between concentration gradient and membrane potential. The voltage calculated from the Nernst equation is the membrane potential that exactly balances the concentration gradient across the membrane.

4. **c.** From the Nernst equation with $[K^+]_o = 10\,mM$ and $[K^+]_i = 100\,mM$, we calculate that the equilibrium potential for K^+ is $-58\,mV$. If the membrane potential equals the equilibrium potential, then the sum of the concentrational and electrical fluxes is zero,

and there is no net movement of the ion across the membrane. $10/100 = 0.1$; $\log(0.1) = -1$; $\mathcal{Z} = +1$ for K^+; thus, $E_K = (58\,\text{mV}/+1)(-1) = -58\,\text{mV}$.

5. **b.** For the Nernst equation for a cation ($\mathcal{Z}$ is positive) to yield a positive equilibrium potential, the logarithm on the right side of the equation must be positive. This requires that the numerator inside the logarithm (the external concentration) must be greater than the denominator (the internal concentration).

6. The Nernst equation for Na^+ is $E_{Na} = (58\,\text{mV}/+1)\log([Na^+]_o/[Na^+]_i)$. With $[Na^+]_o = 120\,\text{mM}$ and $[Na^+]_i = 12\,\text{mM}$ (see Figure 3.1), the equilibrium potential is $+58\,\text{mV}$.

7. The K^+ equilibrium potential is zero. If there is no concentration gradient, then there is no net flux of potassium driven by the concentration gradient. For potassium to be at equilibrium, there must also be no net flux driven by the voltage gradient, which requires that the voltage gradient must be zero.

TOPIC 2: STEADY-STATE MEMBRANE POTENTIAL

KEY POINTS

✓ *What equation is used to calculate the resting membrane potential of a cell whose plasma membrane is permeable to more than one type of ion?*

✓ *Under conditions in which permeant ions are not at equilibrium, how does the cell maintain the differences in ionic composition of the intracellular and extracellular fluids?*

The plasma membrane is simultaneously permeable to several different types of ions. Electrically charged substances cannot diffuse directly through membrane lipids and must either be carried by transport proteins or diffuse through **ion channels**, which are specialized membrane proteins that form an aqueous pore across the membrane. Cell membranes have many kinds of ion channels. Some ion channels allow only Na^+ to cross, whereas others prefer K^+ and still others chloride ions (Cl^-). The membrane may have different permeability to different ions, depending on how many ion channels of each kind are available in the membrane.

In a resting neuron, the permeability to K^+ is about 50 times greater than the permeability to Na^+. Although the permeability to Na^+ is relatively small, it does affect the membrane potential of the neuron. At the resting membrane potential (about $-70\,\text{mV}$), both the concentration gradient and the electrical gradient for Na^+ are inward, and so there is a steady trickle of Na^+ into the cell. The equilibrium potential for K^+ is about $-81\,\text{mV}$. Because of inward flux of Na^+, however, the actual membrane potential is more positive than the K^+ equilibrium potential. Thus, at $-70\,\text{mV}$, neither Na^+—with an equilibrium potential of $+58\,\text{mV}$ (see problem 6 above)—nor K^+ is at equilibrium. There will be a steady efflux of K^+ and a steady influx of Na^+, which causes progressive decline of the ionic gradients across the membrane. Dissipation of the Na^+ and K^+ gradients is prevented by the **sodium-potassium pump** of the plasma membrane. This transport protein uses energy from ATP to pump Na^+ back out of the cell and to pump K^+ into the cell, restoring the ionic gradients.

The resting potential is a nonequilibrium state, with a steady inward flux of Na^+ balanced by a steady outward flux of K^+. When Na^+ influx balances K^+ efflux, the net transfer of charge is zero, and the membrane potential reaches a steady state. The steady-state membrane potential is calculated from the Goldman equation:

$$E_m = \frac{RT}{F}\ln\left(\frac{p_K[K^+]_o + p_{Na}[Na^+]_o}{p_K[K^+]_i + p_{Na}[Na^+]_i}\right) \qquad (3.3)$$

Here, E_m is the membrane potential. The Goldman equation is similar to the Nernst equation, but it simultaneously takes into account all permeant ions. The concentration of each ion inside and outside the cell is scaled according to the permeability of the membrane, p, to that ion. We commonly use a simplified form of the Goldman equation in which the Na^+ permeability is expressed relative to the K^+ permeability, p_K:

$$E_m = 58\,mV \log\left(\frac{[K^+]_o + b[Na^+]_o}{[K^+]_i + b[Na^+]_i}\right) \qquad (3.4)$$

The factor b on the right side of Eq. (3.4) is the ratio of Na^+ to K^+ permeability: $b = p_{Na}/p_K$. As in the simplified form of the Nernst equation [Eq. (3.2)], RT/F has been evaluated at room temperature and base e logarithm has been converted to base 10, yielding the constant 58 mV. In neurons, b is about 0.02 (i.e., p_K is roughly 50 times higher than p_{Na}). For the ionic compositions shown in Figure 3.1, E_m would be approximately -71 mV.

Topic Test 2: Resting Membrane Potential

True/False

1. If the membrane permeability to Na^+ and Cl^- were zero, the Goldman equation would reduce to the Nernst equation for K^+.

2. At the resting membrane potential of a neuron—governed by the Goldman equation—both Na^+ and K^+ are at equilibrium.

3. The sodium-potassium pump maintains the distribution of Na^+ and K^+ across the cell membrane and is thus responsible for maintaining the steady-state membrane potential.

4. If p_{Na} were greater than p_K, the resting membrane potential would be closer to the Na^+ equilibrium potential than to the K^+ equilibrium potential.

Multiple Choice

5. For this question, Na^+ equilibrium potential is $+58$ mV, K^+ equilibrium potential is -80 mV, and the membrane is permeable to both Na^+ and K^+. Ignore chloride.
   a. Neither Na^+ nor K^+ is at equilibrium.
   b. The steady-state membrane potential is between $+58$ mV and -80 mV, at the voltage at which influx of Na^+ exactly balances efflux of K^+.
   c. If Na^+ permeability were increased, leaving K^+ permeability unchanged, the membrane potential would move in the positive direction, toward the Na^+ equilibrium potential.
   d. If both Na^+ permeability and K^+ permeability were doubled, the steady-state membrane potential would not change.
   e. All of the above.

6. Neuron A has a resting membrane potential of -70 mV and neuron B has a resting membrane potential of -75 mV. This could be explained if
   a. p_{Na}/p_K of neuron B is larger than p_{Na}/p_K of neuron A.
   b. p_{Na}/p_K of neuron A is larger than p_{Na}/p_K of neuron B.
   c. p_{Na}/p_K of neuron B is the same as p_{Na}/p_K of neuron A.

Short Answer

7. Calculate the resting membrane potential of a neuron, using Eq. (3.4), if $p_{Na} = p_K$. Assume the ionic conditions shown in Figure 3.1.

Topic Test 2: Answers

1. **True.** In Eq. (3.3), the Na^+ term drops out if p_{Na} is zero. Similarly, Cl^- would not contribute if P_{Cl} were zero. The quantity within the logarithm would then be $[K^+]_o/[K^+]_i$, which is the same as for the Nernst equation. If the cell membrane is permeable to only one type of ion, the membrane potential would be governed exclusively by the equilibrium potential for that ion, given by the Nernst equation.

2. **False.** At the resting potential, neither Na^+ nor K^+ is at equilibrium. The membrane potential cannot simultaneously equal the Na^+ and K^+ equilibrium potentials, which are +58 mV and −81 mV, respectively. Instead, the membrane potential is somewhere in between these equilibrium potentials. There will be a struggle between Na^+ and K^+, with the balance struck when Na^+ influx is exactly equal and opposite to K^+ efflux.

3. **True.** Because neither Na^+ nor K^+ is at equilibrium at the resting potential, Na^+ is continually leaking into the cell and K^+ is continually leaking out. This causes internal Na^+ concentration to rise and internal K^+ concentration to fall, dissipating the gradients. The sodium-potassium pump restores the gradients by transporting Na^+ out of the cell and K^+ into the cell. The pump is a membrane protein that hydrolyzes ATP and uses the released energy to drive the transport across the plasma membrane.

4. **True.** Normally, p_K is much greater than p_{Na}, and so the resting membrane potential is close to the equilibrium potential for the more permeant K^+. However, if p_{Na} were greater than p_K, then the balance between Na^+ and K^+ would be struck closer to the equilibrium potential for the more permeant Na^+.

5. **e.** Statement a is correct because the membrane potential is between the two equilibrium potentials if the membrane is permeable to two ions whose equilibrium potentials are not equal. Thus, neither permeant ion will be at equilibrium. Statement b is correct because the net transfer of charge across the membrane must be zero if the membrane potential is not changing with time. This requires that Na^+ influx must equal K^+ efflux. Statement c is correct because an increase in Na^+ permeability would cause an increase in Na^+ influx, destroying the existing balance between Na^+ influx and K^+ efflux. The influx of positive charge would make the membrane potential more positive. Statement d is correct because the steady-state membrane potential is governed by the ratio of Na^+ to K^+ permeability. If both permeabilities are multiplied by the same factor, the ratio will not change and so the membrane potential will not change.

6. **b.** The resting membrane potential of neuron B is closer to the K^+ equilibrium potential than the resting potential of neuron A. This implies that the ratio of Na^+ to K^+ permeability is smaller in neuron B than in neuron A.

7. $E_m = -2$ mV. If $p_{Na} = p_K$, then the factor b in Eq. (3.4) is 1. With the ionic composition shown in Figure 3.1, the equation becomes

$$E_m = 58\,mV \log\left(\frac{5 + (1 \times (120))}{125 + (1 \times (12))} \right) \approx -2\,mV$$

The resting membrane potential depends on both the ionic permeability of the plasma membrane and on the ionic gradients across the membrane. In certain pathological conditions, reduction in those ionic gradients—and the resulting reduction in membrane potential—can be life threatening. In a heart attack, reduced blood flow (ischemia) interrupts the delivery of oxygen to tissues and the production of ATP falls off. ATP fuels the sodium-potassium pump of the plasma membrane and thus maintains the Na^+ and K^+ gradients across the membrane. Thus, one consequence of ischemia is loss of intracellular K^+ and a corresponding increase in extracellular K^+ (hyperkalemia). From the Nernst equation, doubling external K^+ from 5 mM to 10 mM shifts the K^+ equilibrium potential from about −80 mV to about −63 mV, for example. Because the resting, steady-state membrane potential is near the K^+ equilibrium potential, the resting potential will be less negative when external K^+ rises. The reduced membrane potential is particularly important for cardiac muscle cells, which depend on electrical mechanisms for coordinated pumping of blood. During a heart attack, a vicious cycle can arise in which reduced blood flow leads to hyperkalemia, which in turn interferes further with cardiac function. This cycle can culminate in complete cessation of activity of the heart. The importance of extracellular K^+ for the heart is exploited with lethal intent by the criminal justice systems of some states in the United States, which use K^+ injections to execute condemned prisoners.

Chapter Test

True/False

1. The Nernst equation is used to calculate the membrane potential at which there is no net flux of an ion across the membrane.

2. The equilibrium potential for a monovalent cation is found to be +40 mV. This means that the concentration of this cation in the ECF is less than the concentration in the ICF.

3. Suppose the equilibrium potential for K^+ is −80 mV. If the membrane potential were −90 mV, there would be a net flux of K^+ into the cell.

4. Suppose the concentration of a permeant divalent cation ($z = +2$) is 100 mM in the ECF and 1 mM in the ICF. If the membrane potential were +58 mV, there would be no net flux of the divalent cation across the membrane.

5. Suppose the concentration of Cl^- is less in the ICF than in the ECF. If the membrane potential was 0 mV, there would be no net flux of Cl^- across the membrane.

6. A cell is permeable to K^+ and Cl^-. External K^+ is 10 mM, external Cl^- is 100 mM, internal K^+ is 100 mM, and internal Cl^- is 10 mM. If the membrane potential is −58 mV, there will be no net movement of K^+ across the membrane, but there will be a net influx of Cl^-.

7. A drug is found to depolarize a neuron (i.e., the membrane potential becomes less negative). The drug could do this either by increasing Na^+ permeability or by decreasing K^+ permeability.

8. If p_{Na}/p_K shifts from 0.02 to 20, the membrane potential would shift from near the K^+ equilibrium potential to near the Na^+ equilibrium potential.

9. The resting membrane potential is normally nearer the K^+ equilibrium potential than the Na^+ equilibrium potential because resting K^+ permeability is much less than resting Na^+ permeability.

10. If external K^+ concentration were doubled, leaving all other ion concentrations unchanged, a neuron would depolarize (i.e., the membrane potential would become less negative).

Multiple Choice

11. Suppose the membrane of a cell is permeable to K^+ and we change external K^+ to make $[K]_o = [K]_i$. If the membrane potential of the cell is $0\,mV$,
    a. there will be no net movement of K^+ across the membrane.
    b. there will be a net movement of K^+ into the cell.
    c. there will be a net movement of K^+ out of the cell.
    d. the electrical gradient across the membrane would no longer be balanced by the concentration gradient.

12. Suppose that the concentration of Cl^- inside a cell is greater than the concentration of Cl^- outside the cell. For Cl^- to be at equilibrium, the membrane potential of the cell must be
    a. negative (i.e., the inside of the cell must be negative with respect to the outside).
    b. zero (i.e., there must be no voltage gradient across the membrane).
    c. positive (i.e., the inside of the cell must be positive with respect to the outside).
    d. equal to the equilibrium potential for Na^+ ions so that Na^+ flux balances Cl^- flux.

13. Suppose the Na^+ equilibrium potential is $+58\,mV$, the K^+ equilibrium potential is $-80\,mV$, and E_m is $-70\,mV$. If p_K were increased, without changing p_{Na}, the membrane potential would
    a. remain at $-70\,mV$.
    b. become more negative than $-70\,mV$.
    c. become more positive than $-70\,mV$.
    d. move near the Na^+ equilibrium potential.

14. Suppose a cell has a resting membrane potential of $+30\,mV$ but the Na^+ and K^+ equilibrium potentials have their usual values. This indicates that
    a. p_K is zero.
    b. p_{Na} is less than p_K.
    c. p_{Na} is equal to p_K.
    d. p_{Na} is greater than p_K.

15. The equilibrium potential for an ion
    a. is the membrane potential at which there is no net flux of the ion across the membrane.
    b. is the membrane potential at which the electrical gradient across the membrane exactly balances the concentration gradient for the ion.
    c. can be calculated from the Nernst equation.
    d. all of the above.

Short Answer

16. The _____ equation is used to calculate the equilibrium value of membrane potential for a permeant ion, whereas the _____ equation is used to calculate the steady-state value of membrane potential for a cell whose membrane is permeable to more than one type of ion.

17. If the K^+ equilibrium potential is $-80\,mV$ and the membrane potential is $-60\,mV$, will there be a net flux of K^+ across the membrane? If so, in which direction? Why does K^+ behave as you described?

18. What are the two factors that govern the steady-state membrane potential of a cell, and what roles do the factors play?

19. What is the sodium-potassium pump and why is it necessary for maintenance of the steady-state membrane potential?

Essay

20. Describe the situation for the steady-state membrane potential of a cell, assuming that only Na^+ and K^+ contribute (ignore Cl^-).

Chapter Test Answers

1. **T**   2. **F**   3. **T**   4. **T**   5. **F**   6. **F**   7. **T**   8. **T**   9. **F**   10. **T**   11. **a**   12. **c**
13. **b**   14. **d**   15. **d**

16. Nernst, Goldman

17. Yes. Outward. The equilibrium potential is the membrane voltage at which the electrical gradient just balances the concentration gradient. Because the electrical gradient at $-60\,mV$ is smaller than required to balance the concentration gradient, the flux down the electrical gradient is smaller than the flux down the concentration gradient, and there is a residual outward flux.

18. The two factors are the concentrations of the ions in the ICF and ECF and the permeability of the membrane to the various ions. The concentrations are important because they determine the equilibrium potentials for the ions. The permeability is important because it determines the weight given to a particular ion in determining the steady-state membrane potential.

19. The sodium-potassium pump is a plasma membrane protein that uses ATP to drive movement of Na^+ out of the cell and movement of K^+ into the cell. It is necessary because the steady-state membrane potential is a nonequilibrium condition, with continuous fluxes of Na^+ and K^+ across the membrane. Without the pump, the concentration gradients for Na^+ and K^+ across the membrane would decline.

20. At the steady-state membrane potential, the total flux of charge across the membrane, taking into account all the permeant ions, must be zero. If the total charge flux were not zero, the membrane potential would be changing with time, not steady. If only Na^+ and K^+ contribute, there will be a struggle between Na^+ on the one hand, pushing the membrane potential toward the Na^+ equilibrium potential ($+58\,mV$), and K^+ on the other hand, pushing the membrane potential toward the K^+ equilibrium potential ($-80\,mV$). The balance will be struck when Na^+ influx exactly balances K^+ efflux. K^+ permeability

of the membrane is usually much larger than Na^+ permeability, so that a relatively large flux of K^+ results from a relatively small disparity between the actual value of the membrane potential and the K^+ equilibrium potential. Conversely, because Na^+ permeability is small, a relatively large disparity between the actual membrane potential and the Na^+ equilibrium potential is required to give the same amount of Na^+ influx. Therefore, the balance between Na^+ influx and K^+ efflux will be struck relatively close to the K^+ equilibrium potential and relatively far away from the Na^+ equilibrium potential. The Goldman equation takes into account the equilibrium potentials and the permeabilities of the ions and allows calculation of the membrane potential at which the balance will be struck.

Check Your Performance:

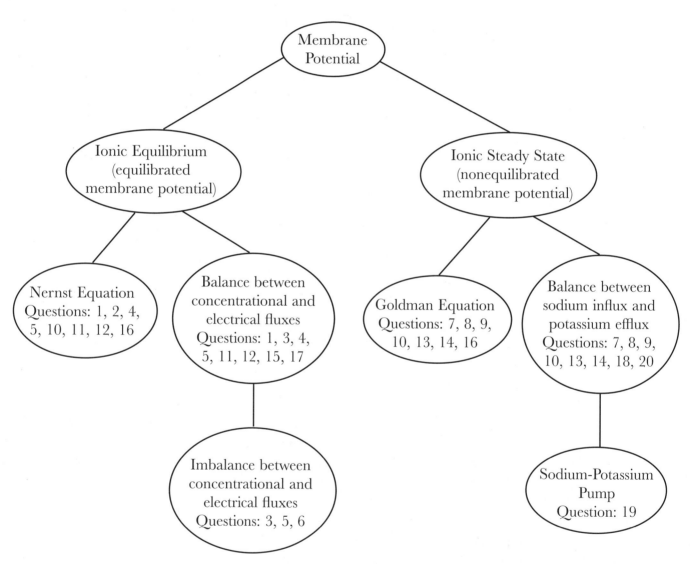

Note the number of questions in each grouping that you got wrong on the chapter test. Identify areas where you need further review and go back to relevant parts of this chapter.

Further help: If you continue to have difficulty with the concepts, a detailed presentation of these topics can be found in chapter 4 of *Neurobiology: Molecules, Cells, and Systems*, published by Blackwell Science, Inc.

CHAPTER 4

The Action Potential

The nervous system transmits information over long distances, using rapidly conducted electrical signals. The long-distance signal of the nervous system is the **action potential**, which is a rapid modulation of the membrane potential of the neuron. The action potential is transmitted from one end of a nerve fiber to the other at speeds as high as 100 m per second. In this chapter, the quantitative principles that govern the membrane potential are used to understand the mechanism of the action potential.

ESSENTIAL BACKGROUND

- **The Nernst equation**
- **The Goldman equation**
- **Electrical voltage and current**

TOPIC 1: CHARACTERISTICS OF THE ACTION POTENTIAL

KEY POINTS

✓ *What does an action potential look like?*

✓ *What stimulus triggers an action potential?*

✓ *How do changes in membrane permeability produce the action potential?*

Figure 4.1A shows the changes in membrane potential during an action potential. The resting membrane potential of the neuron is approximately −70 mV. During the action potential, however, the membrane potential transiently swings in the positive direction, which is called the **depolarizing phase** (or the **upstroke**) of the action potential.

At the peak of the action potential, the membrane potential actually reverses sign briefly, so that the inside of the cell becomes positive with respect to the outside. This sign reversal is called the **overshoot** of the action potential. The membrane potential then rapidly returns to negative values, which is called the **repolarizing phase** (or the **downstroke**). When the membrane potential returns toward rest, it may briefly become more negative than normal, a period called the **undershoot**. The stimulus that initiates an action potential is reduction in the membrane potential—that is, a depolarization of the membrane. Normally, the depolarization that triggers an action potential is produced by some external stimulus, such as the reception of a sensory stimulus or the release of chemical transmitter at a synapse (see Chapter 5).

The key to understanding the origin of the action potential lies in the discussion in Chapter 3 of the factors that govern the steady-state membrane potential. Recall that the resting membrane

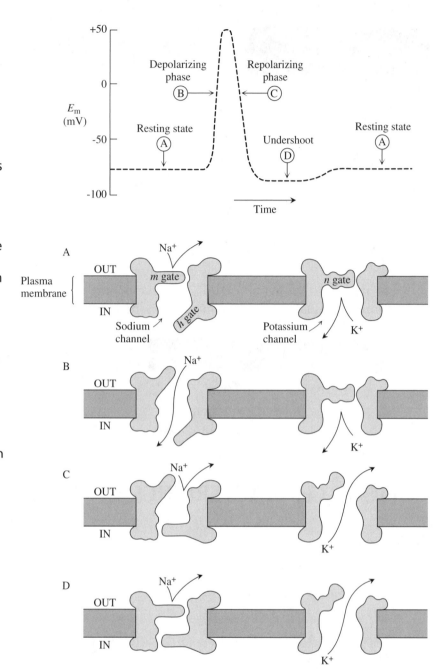

Figure 4.1 The state of voltage-sensitive sodium and potassium channels at various times during an action potential in a neuron. A. At rest, neither channel is in a conducting state. B. During the depolarizing phase of the action potential, sodium channels open, but potassium channels have not had time to respond to the depolarization. C. During the repolarizing phase, sodium permeability begins to return to its resting level as h gates have sufficient time to respond to the depolarizing phase. At about the same time, potassium channels respond to the depolarization and open. D. During the undershoot, sodium permeability has returned to its usual low level; potassium permeability, however, remains elevated because n gates respond slowly to the repolarization of the membrane. The resting state of the membrane is restored as h gates and n gates have time to respond to the repolarization.

potential is somewhere between the K^+ and Na^+ equilibrium potentials, E_K and E_{Na}, and that the ratio of Na^+ to K^+ permeability, p_{Na}/p_K, determines the exact balance point. Resting p_{Na}/p_K is approximately 0.02, and the resting membrane potential is relatively close to E_K. However, if p_{Na} increased about 1,000-fold, leaving p_K unchanged, p_{Na}/p_K would increase to 20, and from the Goldman equation, we calculate that the membrane potential would shift from about $-70\,mV$ to about $+50\,mV$. If p_{Na}/p_K then returns to 0.02, the membrane potential moves back near E_K. This sequence of permeability changes underlies the action potential.

The depolarizing phase of the action potential is fueled by a large increase in p_{Na}, which changes p_{Na}/p_K from 0.02 to about 20 and drives membrane potential near E_{Na}. This explains the over-shoot of the action potential: The membrane potential at the peak of the action potential reverses sign because E_{Na} is positive. The repolarizing phase of the action potential is driven by two factors. First, the increase in p_{Na} is inherently transient, so that p_{Na} returns to its low resting level. Second, there is a delayed increase in p_K, which makes p_K even higher than its resting level. The combination of returning p_{Na} to rest and increasing p_K makes p_{Na}/p_K even smaller than 0.02, driving the membrane potential even closer to E_K than normal. The fall of p_{Na}/p_K below 0.02 accounts for the undershoot of the action potential. Ultimately, p_K returns to rest, p_{Na}/p_K is restored to 0.02, and the membrane potential is once again back to the resting state (**Figure 4.1A**).

Topic Test 1: Characteristics of the Action Potential

True/False

1. At the peak of the action potential, the membrane potential becomes positive because p_K is much greater than p_{Na}, driving the membrane potential near E_K.

2. The undershoot of the action potential corresponds to the time when p_{Na} has returned to its resting level, whereas p_K remains elevated.

3. The resting potential of a neuron is governed by p_{Na}/p_K, as specified by the Goldman equation, but during an action potential, the membrane potential is no longer determined by p_{Na}/p_K.

Multiple Choice

4. During an action potential,
   a. the membrane potential at the peak of the action potential becomes transiently positive because the p_K goes to zero whereas the p_{Na} remains at its resting level. This drives the membrane potential near E_K.
   b. a large increase in p_{Na} produces the upstroke of the action potential, driving the membrane potential near E_{Na}.
   c. the downstroke is produced by a large decrease in p_K, which causes the membrane potential to move away from E_{Na}.
   d. none of the above.

5. At the peak of the action potential, the membrane potential is closest to the equilibrium potential for which ion?
   a. K^+
   b. water
   c. Na^+
   d. protein

Short Answer

6. Describe the permeability changes that underlie the depolarizing phase and the repolarizing phase of the action potential.

Topic Test 1: Answers

1. **False.** The membrane potential at the peak of the action potential does indeed become positive, but this is because p_{Na} increases dramatically, becoming much greater than p_K, driving the membrane potential near E_{Na}.

2. **True.** During the undershoot, membrane potential transiently moves closer to E_K, making the potential more negative than usual. This occurs because p_K is higher than it is at rest, whereas p_{Na} returns to rest, making p_{Na}/p_K smaller than normal.

3. **False.** At all times, the membrane potential is governed by the relative permeabilities of the permeant ions, as specified by the Goldman equation. For a neuron, the relevant ions are Na^+ and K^+, and the permeability ratio p_{Na}/p_K determines the membrane potential. The action potential arises because of changes in p_{Na}/p_K, not because of some alteration in the nature of the relation between membrane potential and ionic permeability.

4. **b.** The depolarizing phase of the action potential occurs because of a dramatic increase in p_{Na}, which makes p_{Na}/p_K approximately 20. This change in p_{Na}/p_K causes the membrane potential to move near E_{Na}. Therefore, choice b is correct, and choice a is incorrect. Choice c is incorrect, because the downstroke of the action potential is in fact produced by two factors: p_{Na} returns to its resting value and p_K increases.

5. **c.** At the peak of the action potential, the membrane is more permeable to Na^+ than to other ions, and so the membrane potential approaches E_{Na}. E_K and E_{Cl} are negative (Chapter 3). Water is not an ion, and the membrane is not permeable to protein molecules.

6. The depolarizing phase is produced by an increase in p_{Na}. The repolarizing phase is produced by the return of p_{Na} to its low resting value and by an increase in p_K.

TOPIC 2: VOLTAGE-DEPENDENT ION CHANNELS

KEY POINTS

✓ *What ion channels produce the changes in permeability during the action potential?*

✓ *How do the gates of the ion channel change in response to changes in membrane potential?*

Ions cross the membrane through transmembrane protein molecules called ion channels. The changes in p_{Na} and p_K during an action potential are produced by special ion channels that open and close in response to changes in membrane potential. The increase in p_{Na} during the depolarizing phase is produced by **voltage-dependent sodium channels**. These Na^+ channels are closed as long as the membrane potential is near its normal negative level and open upon depolarization. The increase in p_{Na} upon depolarization causes an increased flux of Na^+ into the cell, leading to greater depolarization and hence to the opening of still more voltage-dependent Na^+ channels. Thus, the response to depolarization is explosive, producing a large increase in p_{Na} and driving the membrane potential near E_{Na}.

The opening of the Na^+ channels upon depolarization is transient, however, and the channels close again within 1 or 2 milliseconds after depolarization. The Na^+ channel behaves as though the flow of Na^+ is controlled by two independent gates (**Figure 4.1B**). One gate, the m gate, remains closed when the membrane potential is equal to or more negative than the usual resting potential. This gate prevents Na^+ from entering the channel at the resting potential. The other gate, the h gate, is open at the usual resting membrane potential. Both gates respond to depolarization, but with different speeds and in opposite directions. The m gate opens rapidly in response to depolarization; the h gate closes slowly in response to depolarization. The rapid opening of the m gate produces the upstroke of the action potential, and the delayed closing of the h gate produces the subsequent decline of p_{Na} back to its resting level. Because opening of the m gates opens the channel, the m gate is also called the activation gate. Because closing of the h gate closes the channel, the h gate is also called the inactivation gate.

In addition to voltage-dependent Na^+ channels, neurons also have voltage-dependent K^+ channels. Like the sodium channel m gates, the gates of the K^+ channels (called n gates) are closed at the resting membrane potential and open upon depolarization, allowing K^+ to move through the channel. The n gates open slowly upon depolarization, causing the delayed increase in p_K that is partly responsible for the repolarizing phase. Unlike the Na^+ channel, the K^+ channel has no inactivation gate. Instead, the channel remains open as long as depolarization is maintained and closes only when membrane potential returns to rest.

Figure 4.1B summarizes the behavior of the voltage-dependent Na^+ and K^+ channels during the action potential. The undershoot of the action potential corresponds to the period when the potassium channel n gates remain open after repolarization. The sodium channel h gates remain closed for a period after repolarization. During this period, called the refractory period, the neuron cannot fire another action potential, even if it receives another depolarizing stimulus, because the channels will not conduct Na^+ while the h gates are closed, even if the m gates open again in response to depolarization.

Topic Test 2: Voltage-Dependent Ion Channels

True/False

1. During an action potential, p_{Na} first increases, as m gates open, then decreases as h gates close.

2. An action potential propagates without decrement from one end of a nerve fiber to the other because the large depolarization produced by the action potential in one region of the fiber brings the next segment of the fiber above threshold for triggering a new action potential in the next segment.

3. During the repolarizing phase of the action potential, the two mechanisms of repolarization are the opening of the sodium channel m gates and the closing of potassium channel n gates.

Multiple Choice

4. Select the correct statement.
   a. The refractory period corresponds to the time after an action potential when h gates of Na^+ channels remain closed.
   b. The undershoot of the action potential corresponds to the time when p_{Na} has returned to its resting level while p_K remains elevated.
   c. In the resting state, both the voltage-dependent Na^+ channels and the voltage-dependent K^+ are closed.
   d. all of the above.
   e. a and b are correct, but not c.

5. For the following statements, suppose we take control of the membrane potential of a neuron. We step the membrane potential from $-70\,mV$ to $0\,mV$ and hold it there.

a. The sodium-channel m gates open rapidly and remain open as long as membrane potential remains at 0 mV. After the h gates have time to close, the h gates will remain closed as long as membrane potential remains at 0 mV.

b. p_K will not increase initially. With a delay, p_K increases, as potassium-channel n gates open in response to the depolarization.

c. After the n gates have time to respond to the depolarization, the K$^+$ channel n gates will remain open for only a brief time as we maintain the membrane potential at 0 mV. p_K therefore declines back to rest during a maintained depolarization.

d. All of the above.

e. a and b are correct, but not c.

Short Answer

6. Briefly summarize responses of the gates of voltage-dependent Na$^+$ and K$^+$ channels to depolarization. Include a description of the speed of response of the gates.

Topic Test 2: Answers

1. **True.** The m gates of Na$^+$ channels open rapidly to depolarization, increasing p_{Na} and producing the depolarizing phase. Then, with a delay, the h gates close, and p_{Na} declines.

2. **True.** The trigger for an action potential is depolarization, and the action potential is itself a large depolarization. Normally, an action potential is triggered in a neuron by some external stimulus. Once triggered, the action potential propagates along the axon from one end to the other. The depolarization produced by the action potential in each location causes Na$^+$ channels to open in the neighboring segment of the axon, producing an action potential at that location. The resulting depolarization in turn activates Na$^+$ channels in the next segment of the axon. The result is that a wave of depolarization propagates progressively along the axon. The action potential propagates in only one direction, because the region just traversed by the action potential is in the refractory period.

3. **False.** Repolarization occurs because p_{Na} declines (as h gates close) and p_K increases (as n gates open).

4. **d.** Statements a through c are all correct.

5. **e.** Statement a is correct because during a sustained depolarization, the m gates will remain open and the h gates will remain closed. The gates will not revert to their resting state (m gates closed and h gates open) until the membrane potential is restored to its negative resting value. Statement b is correct because the K$^+$ channel n gates open slowly upon depolarization. Therefore, immediately after depolarization, n gates remain closed and the p_K remains at its resting level. With time, n gates open and p_K increases. Statement c is false because K$^+$ channels are controlled by a single gate, which remains open until the membrane potential returns to the negative resting value. Therefore, during sustained depolarization, p_K increases and remains elevated as long as depolarization is sustained.

6. Upon depolarization, the Na$^+$ channel m gate opens rapidly, the Na$^+$ channel h gate closes slowly, and the K$^+$ channel n gate opens slowly.

Action potentials transmit information along the fibers that connect neurons with each other or with their sensory or motor targets. If conduction of action potentials is prevented, the information conveyed by the neuron is blocked. Blockade of action potentials is frequently used in medicine to produce local anesthesia. Many local anesthetics act by blocking voltage-dependent Na^+ channels. Commonly used local anesthetics that act in this way are procaine, lidocaine, and tetracaine. Interestingly, the first local anesthetic of this type to be described and used clinically was cocaine, a drug better known for its other pharmacological actions in the central nervous system and for its addictive properties. Injection of local anesthetic near a peripheral nerve blocks action potentials in all sensory and motor nerve fibers within the nerve, producing loss of all sensation and motor paralysis in the area of the body supplied by that nerve. Naturally occurring biological toxins are also known to block the Na^+ channels of nerve membranes, including the puffer fish toxin tetrodotoxin and saxitoxin. Saxitoxin is produced by marine dinoflagellates and is responsible for paralytic shellfish poisoning in humans who eat shellfish that have ingested the poisonous dinoflagellates ("red tide" organisms).

Chapter Test

True/False

1. The upstroke of the action potential is an explosive process in which depolarization opens Na^+ channels, which increases Na^+ influx, which in turn produces further depolarization.

2. At the peak of the action potential, p_{Na} is greater than p_K.

3. During the undershoot of the action potential, the ratio p_{Na}/p_K is greater than normal.

4. Upon depolarization, the Na^+ channel h gates open with a delay, and upon repolarization the h gates close again with a delay.

5. Na^+ channel m gates respond rapidly to both depolarization and repolarization, whereas Na^+ channel h gates and K^+ channel n gates respond slowly to both depolarization and repolarization.

Multiple Choice

6. During the repolarizing phase of the action potential,
   a. p_K is decreasing because K^+ channel n gates are closing.
   b. p_{Na} is increasing because m gates of the sodium channels are reopening as membrane potential becomes more negative.
   c. Na^+ channel h gates are closing, causing p_{Na} to decline.
   d. p_K has returned to its normal resting value.

7. During the depolarizing phase of the action potential,
   a. p_K is decreasing, driving the membrane potential in the positive direction.
   b. p_{Na} is increasing, driving the membrane potential toward E_{Na}.
   c. p_{Na}/p_K is the same as in the resting state.
   d. K^+ is flowing into the cell to produce the depolarization.

8. During an action potential, p_{Na}
   a. first increases, then decreases back to the resting level.
   b. first decreases, then increases back to the resting level.
   c. does not change.
   d. is always less than p_K.

9. In voltage-dependent Na^+ channels,
   a. the m gates open rapidly upon depolarization, whereas the h gates close slowly upon depolarization.
   b. the h gates open slowly upon depolarization, whereas the m gates open rapidly upon depolarization.
   c. the m gates open rapidly upon depolarization but close slowly upon repolarization.
   d. sodium ions can flow through the channel if either the m gate is open or the h gate is open but not if both are open.

10. In voltage-dependent K^+ channels,
    a. the n gates open slowly upon depolarization but close rapidly upon repolarization.
    b. the n gates close upon depolarization and reopen upon repolarization.
    c. the n gates are open in the resting state.
    d. the n gates open slowly upon depolarization and remain open for a period after repolarization.

Short Answer

11. The ___ gate of the _____ channel opens rapidly upon depolarization.
12. The ___ gate of the _____ channel opens slowly upon depolarization.
13. The ___ gate of the _____ channel closes slowly upon depolarization.
14. What are the states (closed or open) of the gates of the voltage-dependent Na^+ and K^+ channels in the normal resting state?
15. When an action potential travels along a nerve fiber and reaches the end, why doesn't it then reverse direction and travel back in the opposite direction?

Essay

16. How does the action potential propagate along a nerve fiber?
17. Describe the sequence of channel gating in voltage-dependent Na^+ and K^+ channels in the various phases of the action potential.

Chapter Test Answers

1. **T** 2. **T** 3. **F** 4. **F** 5. **T** 6. **c** 7. **b** 8. **a** 9. **a** 10. **d**

11. m, sodium
12. n, potassium
13. h, sodium
14. In the resting state, h gates are open, m gates are closed, and n gates are closed.
15. The refractory period after an action potential prevents an action potential from reversing and traveling back over the region it has just traversed. After an action potential, the Na^+ channel h gates remain closed for a period and prevent the immediate occurrence of another action potential.
16. The depolarization produced by the action potential in one region of the fiber opens voltage-dependent Na^+ channels in the adjacent region of the fiber. Opening of Na^+

channels further depolarizes the adjacent region, driving the membrane potential at that point near E_{Na}. In other words, a new action potential is triggered in the adjacent region. The depolarization produced by this action potential then opens Na^+ channels in the next segment of the fiber. In this manner, the channel gating events producing the action potential sweep along the fiber from one end to the other, recreating the action potential at each succeeding part of the fiber.

17. In the resting state, voltage-dependent Na^+ channels are closed, because m gates are closed, even though h gates are open. Voltage-dependent K^+ channels are also closed in the resting state, because n gates are closed. When the neuron receives a depolarizing stimulus, m gates rapidly open and Na^+ channels begin to conduct Na^+ ions. The influx of Na^+ causes still more depolarization, which opens more Na^+ channels. This produces rapid depolarization during the depolarizing phase of the action potential. Depolarization also opens n gates of K^+ channels, but they do so slowly. At about the same time that the n gates are opening, the Na^+ channel h gates close in response to depolarization. Therefore, p_{Na} declines and p_K increases, both of which drive membrane potential back toward the resting level. In response to repolarization, Na^+ channel m gates close rapidly. Given sufficient time, Na^+ channel h gates respond to repolarization by reopening, and K^+ channel n gates respond to repolarization by closing. However, both the h gates and n gates respond slowly to repolarization. Therefore, for a brief period after the action potential, h gates remain closed, accounting for the refractory period. In addition, although K^+ channel n gates remain open, p_K is elevated, which drives membrane potential even closer to E_K than at rest. This accounts for the undershoot of the action potential.

Check Your Performance:

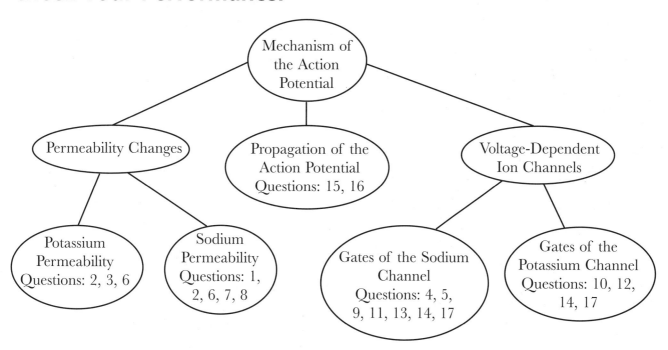

Note the number of questions in each grouping that you got wrong on the chapter test. Identify areas where you need further review and go back to relevant parts of this chapter.

Further help: If you continue to have difficulty with the concepts, a detailed presentation of these topics can be found in chapter 5 of *Neurobiology: Molecules, Cells, and Systems*, published by Blackwell Science, Inc.

Synaptic Transmission

CHAPTER 5

Activity is transmitted from one neuron to another or from a motor neuron to its motor target at a contact point called a **synapse**. Synaptic transmission is either electrical or chemical. At a chemical synapse, the membranes of the input cell (the presynaptic cell) and the receiving cell (the postsynaptic cell) come close to each other but are separated by a small gap, the **synaptic cleft**. An action potential in the presynaptic cell causes the release of a chemical neurotransmitter substance, which diffuses across the synaptic cleft and alters the membrane potential of the postsynaptic cell. At an electrical synapse, changes in membrane potential of the presynaptic cell spread directly to the postsynaptic cell without the intervention of an intermediary chemical. This chapter focuses on chemical synapses.

ESSENTIAL BACKGROUND

- **The Nernst equation**
- **The Goldman equation**
- **Electrical voltage and current**
- **Action potential and voltage-dependent channels**

TOPIC 1: PRESYNAPTIC EVENTS IN SYNAPTIC TRANSMISSION

KEY POINTS

✓ *What aspect of a presynaptic action potential triggers neurotransmitter release?*

✓ *How is presynaptic depolarization linked to neurotransmitter release?*

✓ *What is the mechanism of neurotransmitter release?*

The presynaptic component of the synapse is the **synaptic terminal**, which contains machinery for the release of neurotransmitter. Neurotransmitter release is triggered by depolarization of the synaptic terminal, which is normally produced by a presynaptic action potential. **Figure 5.1** summarizes the linkage between depolarization and neurotransmitter release. Depolarization opens voltage-dependent calcium channels in the synaptic terminal, and calcium ions (Ca^{2+}) flow into the terminal. This influx increases intracellular Ca^{2+}, which is sensed by Ca^{2+} sensor molecules that trigger neurotransmitter release. Neurotransmitter is stored in membrane-bound structures called **synaptic vesicles**. Each synaptic vesicle contains several thousand molecules of neurotransmitter. To release neurotransmitter into the synaptic cleft, the vesicle membrane

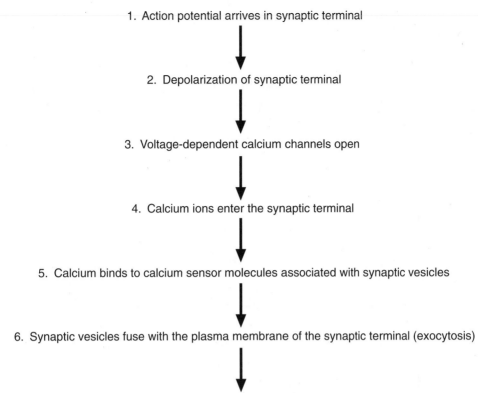

1. Action potential arrives in synaptic terminal

2. Depolarization of synaptic terminal

3. Voltage-dependent calcium channels open

4. Calcium ions enter the synaptic terminal

5. Calcium binds to calcium sensor molecules associated with synaptic vesicles

6. Synaptic vesicles fuse with the plasma membrane of the synaptic terminal (exocytosis)

7. Contents of synaptic vesicle (several thousand molecules of neurotransmitter) are released into the extracellular space (synaptic cleft)

Figure 5.1 The sequence of presynaptic events at a chemical synapse.

must fuse with the plasma membrane of the terminal. Thus, neurotransmitter release occurs by exocytosis.

Topic Test 1: Presynaptic Events in Synaptic Transmission

True/False

1. When voltage-dependent Ca^{2+} channels open in the synaptic terminal, Ca^{2+} flows out of the terminal.

2. Depolarization produced by a presynaptic action potential is the stimulus that opens Ca^{2+} channels in the synaptic terminal.

3. Neurotransmitter molecules are released from the synaptic terminal a single molecule at a time.

Multiple Choice

4. During an action potential in the synaptic terminal,
   a. voltage-dependent Ca^{2+} channels open during the depolarization produced by the action potential.
   b. Ca^{2+} flows into the synaptic terminal and triggers neurotransmitter release.
   c. neurotransmitter molecules are released from the synaptic terminal via exocytosis.

d. all of the above.

e. none of the above.

5. Synaptic vesicles

    a. release Ca^{2+} into the intracellular fluid of the terminal in response to depolarization.

    b. are the part of the plasma membrane of the synaptic terminal where voltage-dependent Ca^{2+} channels are located.

    c. are another name for synaptic terminals.

    d. all of the above.

    e. none of the above.

Short Answer

6. During a presynaptic action potential, voltage-dependent _____ channels open, allowing influx of _____ ions.

7. Neurotransmitter release occurs by means of _____, as the membranes of synaptic _____ fuse with the plasma membrane, releasing (circle correct number: one, tens, hundreds, thousands, millions, billions) neurotransmitter molecule(s) into the _____.

Topic Test 1: Answers

1. **False.** Ca^{2+} actually flows into the terminal when Ca^{2+} channels open. The direction of Ca^{2+} flux is determined by the Ca^{2+} concentration inside and outside the cell and by the transmembrane voltage. External Ca^{2+} concentration is low (about 1 mM), but the intracellular concentration is much lower (about 0.1 µM). The concentration gradient therefore drives inward movement of Ca^{2+} through open Ca^{2+} channels. The electrical gradient also drives Ca^{2+} influx. From the Nernst equation, for $[Ca^{2+}]_o = 10^{-3}$ M and $[Ca^{2+}]_i = 10^{-7}$ M, the equilibrium potential is +116 mV ($z = +2$ for calcium ions), and so even at the peak of the action potential, the membrane potential is more negative than the equilibrium potential. So, both concentration and electrical gradients drive Ca^{2+} into the cell.

2. **True.** The Ca^{2+} channels of the synaptic terminal are voltage-dependent channels similar to the voltage-dependent Na^+ and K^+ channels of the action potential. Depolarization opens Ca^{2+} channels, which close again upon repolarization. So, the channels open during the depolarization produced by the action potential and close again when the membrane potential returns to its resting negative level.

3. **False.** Neurotransmitter molecules are stored in synaptic vesicles, which contain several thousand molecules of neurotransmitter. Because release occurs via exocytosis, the entire contents of the synaptic vesicle are released at once. Therefore, neurotransmitter release occurs in irreducible packets (called quanta), corresponding to the entire contents of a single synaptic vesicle.

4. **d.** Depolarization during the presynaptic action potential opens voltage-dependent Ca^{2+} channels. The resulting Ca^{2+} influx triggers neurotransmitter release via exocytosis, as the membranes of synaptic vesicles fuse with the plasma membrane, releasing the neurotransmitter content of the vesicle into the synaptic cleft.

5. **e.** Synaptic vesicles are actually intracellular membrane-bound storage sites for neurotransmitter.

6. **calcium, calcium.** Depolarization during the action potential opens voltage-dependent Ca^{2+} channels, and the influx of Ca^{2+} ions triggers neurotransmitter release.

7. exocytosis, vesicles, thousands, synaptic cleft (or extracellular space). The fusion of the membrane of the synaptic vesicle with the plasma membrane is a form of exocytosis. The thousands of neurotransmitter molecules contained inside the synaptic vesicle are then released into the extracellular space (the synaptic cleft).

TOPIC 2: POSTSYNAPTIC EVENTS IN SYNAPTIC TRANSMISSION

KEY POINTS

✓ *How does the postsynaptic cell detect the presence of released neurotransmitter?*

✓ *How does the neurotransmitter substance affect the membrane potential of the postsynaptic cell?*

✓ *What is the difference between an excitatory synapse and an inhibitory synapse?*

The sequence of postsynaptic events triggered by neurotransmitter release is summarized in **Figure 5.2**. Neurotransmitter molecules are released from the presynaptic terminal and diffuse across the narrow gap, separating the presynaptic and postsynaptic cells. The membrane of the postsynaptic cell has a high density of **neurotransmitter receptor molecules**, which are transmembrane proteins whose extracellular region includes a binding site that recognizes the neurotransmitter molecule.

Many different neurotransmitter substances are released at different synapses. Most neurotransmitter substances are small molecules, including amino acids (e.g., glutamate and glycine) and simple derivatives of amino acids (norepinephrine and dopamine derived from tyrosine, serotonin derived from tryptamine, and GABA derived from glutamate). Another common neurotransmitter is acetylcholine, which is formed by an ester linkage between acetate and choline. Correspondingly, many different types of receptor proteins exist, each specialized for a particular neurotransmitter (e.g., glutamate receptors, acetylcholine receptors, etc.).

Binding of neurotransmitter to receptor molecules changes the ionic permeability of the postsynaptic cell and alters the membrane potential of the postsynaptic cell. Linkage between the receptor molecule and the change in permeability may be either direct or indirect. In the case of direct action, the ion channel is part of the receptor itself, and the channel opens when neurotransmitter binds. In the case of indirect action, the receptor molecule is not itself an ion channel but instead sets in motion a series of events culminating in opening or closing postsynaptic ion channels.

Many indirectly acting neurotransmitter receptors are **G protein-coupled receptors**, which activate signal molecules called **GTP-binding proteins** (or **G proteins**). G proteins modulate the activity of enzymes that synthesize or degrade intracellular second messenger substances, and the second messenger then opens or closes ion channels, either by directly binding to them or by activating an enzyme that acts on the ion channels. A given neurotransmitter substance may have postsynaptic receptors of both types. For instance, some acetylcholine receptors (nicotinic

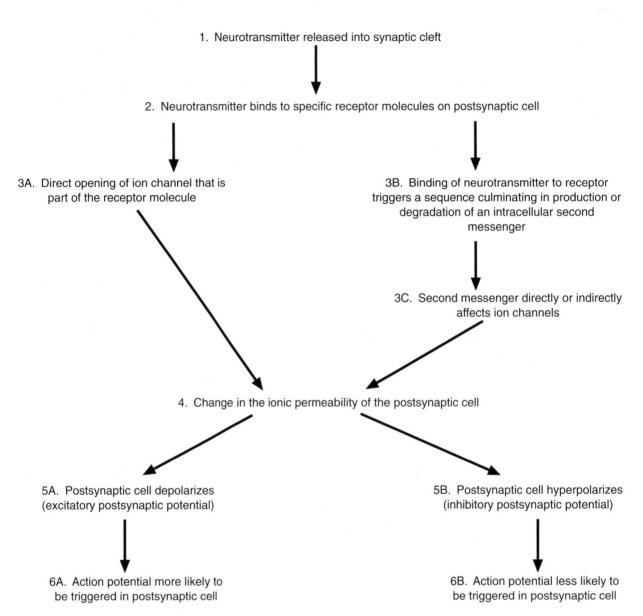

1. Neurotransmitter released into synaptic cleft

2. Neurotransmitter binds to specific receptor molecules on postsynaptic cell

3A. Direct opening of ion channel that is part of the receptor molecule

3B. Binding of neurotransmitter to receptor triggers a sequence culminating in production or degradation of an intracellular second messenger

3C. Second messenger directly or indirectly affects ion channels

4. Change in the ionic permeability of the postsynaptic cell

5A. Postsynaptic cell depolarizes (excitatory postsynaptic potential)

5B. Postsynaptic cell hyperpolarizes (inhibitory postsynaptic potential)

6A. Action potential more likely to be triggered in postsynaptic cell

6B. Action potential less likely to be triggered in postsynaptic cell

Figure 5.2 The sequence of postsynaptic events at a chemical synapse.

receptors) are of the direct type, and other acetylcholine receptors (muscarinic receptors) are coupled indirectly to ion channels through G proteins.

The effect of a neurotransmitter on the postsynaptic cell depends on what ions can cross through the affected ion channels. Opening Na^+ channels increases p_{Na} and depolarizes the postsynaptic cell, whereas opening K^+ channels or Cl^- channels hyperpolarizes the postsynaptic cell. Note that closing Na^+ channels also hyperpolarizes the postsynaptic cell and that closing K^+ or Cl^- channels depolarizes the cell. Depolarization promotes firing action potentials in the postsynaptic cell, and thus a synapse where the neurotransmitter depolarizes the cell is an excitatory synapse.

Depolarization produced by activation of an excitatory synapse is called an **excitatory postsynaptic potential** (epsp). Conversely, a synapse at which the neurotransmitter hyperpolarizes the postsynaptic cell tends to prevent action potentials and is an inhibitory synapse. Hyperpolarization produced by activation of an inhibitory synapse is called an **inhibitory postsynaptic potential** (ipsp). At a given instant, all of the ipsps and epsps from all active synaptic inputs to a neuron sum together to determine whether the neuron fires an action potential.

Topic Test 2: Postsynaptic Events in Synaptic Transmission

True/False

1. An excitatory postsynaptic potential could occur if neurotransmitter opens channels that are equally permeable to Na^+ and K^+.

2. An inhibitory postsynaptic potential could occur if neurotransmitter opens either K^+ or Cl^- channels.

3. Whether a particular synapse is excitatory or inhibitory is determined by the identity of the neurotransmitter released by the presynaptic cell.

Multiple Choice

4. The postsynaptic effect of neurotransmitter at a synapse
   a. is mediated by specific receptor molecules in the postsynaptic membrane.
   b. produces either excitation or inhibition of the postsynaptic cell.
   c. may be mediated indirectly via an intracellular second messenger.
   d. may result from direct opening of an ion channel that is part of the receptor molecule.
   e. all of the above.

5. An excitatory postsynaptic potential could result from
   a. closing sodium channels.
   b. opening potassium channels.
   c. closing nonspecific cation channels, thus reducing p_{Na} and p_K by the same amount.
   d. closing potassium channels.
   e. all of the above are correct.

Short Essay

6. Describe the linkage between G protein-coupled receptors and ion channels. Give an example.

Topic Test 2: Answers

1. **True.** Many epsps are in fact produced by opening nonspecific cation channels that are equally permeable to Na^+ and K^+. For example, the synapse between motor neurons and muscle cells (the neuromuscular junction) works this way. In the resting state, p_K is about 50 times greater than p_{Na} ($p_{Na}/p_K = 0.02$ in the Goldman equation). If p_{Na} is 1 unit of permeability, then p_K at rest is 50 units. If the neurotransmitter increases both p_{Na} and p_K by 50 units by opening channels that are equally permeable to Na^+ and K^+, then p_{Na} would increase from 1 unit to 51 units, but p_K would increase from 50 units to 100 units. The ratio p_{Na}/p_K then increases from 0.02 to 0.51, and the membrane potential calculated from the Goldman equation is substantially more positive.

2. **True.** The equilibrium potentials for Cl^- and K^+ are more negative than the resting potential. So, opening a K^+ channel allows efflux of positive charge and opening a Cl^- channel allows influx of negative charge. In both cases, membrane potential moves toward to equilibrium potential for the ion, and the neuron hyperpolarizes.

3. **False.** Whether a neurotransmitter produces excitation or inhibition is determined by the permeability characteristics of the ion channels affected by the neurotransmitter. A given neurotransmitter substance may be excitatory at one synapse but inhibitory at a different synapse.

4. **e.** Neurotransmitter molecules released from the synaptic terminal combine with specific receptor molecules in the postsynaptic cell. The receptor molecule affects ionic permeability of the postsynaptic cell, either by opening an ion channel that is part of the receptor protein or by an intracellular second messenger that in turn affects postsynaptic channels.

5. **d.** Reducing p_K increases p_{Na}/p_K, which depolarizes the postsynaptic cell and produces an epsp. All other changes in postsynaptic permeability in question would actually decrease p_{Na}/p_K, which hyperpolarizes the postsynaptic cell and produces an ipsp.

6. When neurotransmitter binds to the receptor, the receptor interacts with and activates an intracellular GTP-binding protein (G protein). The G protein then activates an enzyme that synthesizes or degrades an intracellular second messenger. The second messenger then either directly interacts with ion channels or stimulates an enzyme that modulates the activity of ion channels. An example is the effect of a particular subtype of beta-adrenergic receptors on voltage-dependent Ca^{2+} channels. Beta-adrenergic receptors are a type of receptor for the neurotransmitter norepinephrine (also called noradrenaline). When norepinephrine binds, the receptor activates a G protein that stimulates the enzyme adenylyl cyclase, which synthesizes the intracellular second messenger, cyclic AMP. So, activation of beta-adrenergic receptors increases cyclic AMP in the postsynaptic cell. Cyclic AMP activates another enzyme, protein kinase A, which phosphorylates voltage-dependent Ca^{2+} channels. Phosphorylation of the Ca^{2+} channel promotes opening of the channel in response to depolarization.

IN THE CLINIC

Neurotransmitters and their postsynaptic receptors are fundamental for the nervous system. Therefore, many drugs used clinically exert their effects by mimicking or blocking the action of neurotransmitters. Numerous tranquilizers, stimulants, analgesics, antipsychotic, and sedative drugs act either presynaptically or postsynaptically to affect neurotransmission in the central nervous system. Pharmacologists are exploiting the wide variety of different neurotransmitter receptors found at different synapses in the nervous system to devise drugs that target selective aspects of nervous system function. Drugs that affect neurotransmission in the peripheral nervous system are also important clinically. Drugs acting at norepinephrine synapses in the autonomic nervous system are useful in controlling high blood pressure and increasing cardiac output. Many biological toxins also target synaptic transmission. For example, the fish-eating marine snail, *Conus geographicus*, paralyzes its prey with a venom containing omega-conotoxin, which blocks voltage-dependent Ca^{2+} channels of synaptic terminals. Another example is the poisonous snake, *Bungarus multicinctus*, whose venom includes alpha-bungarotoxin, which binds to and blocks the postsynaptic acetylcholine receptors at the neuromuscular junction.

Chapter Test

True/False

1. Removing Ca^{2+} from the extracellular fluid blocks release of excitatory neurotransmitters but has no effect on the release of inhibitory neurotransmitters.

2. Synaptic vesicles fuse with the plasma membrane of the postsynaptic cell at a synapse.

3. Depolarization produced by an action potential opens voltage-dependent Ca^{2+} channels in the synaptic terminal.

4. A neurotransmitter that opens K^+ channels in the postsynaptic cell would produce an ipsp.

5. Neurotransmitter is released from the synaptic terminal in multimolecular packets corresponding to the amount of neurotransmitter that can flow through a single Ca^{2+} channel.

Multiple Choice

6. Neurotransmitter receptor molecules
   a. are located in the synaptic terminal of the presynaptic neuron.
   b. are transmembrane proteins that bind neurotransmitter and either directly or indirectly affect ion channels.
   c. are responsible for fusion of synaptic vesicles with the plasma membrane of the synaptic terminal.
   d. allow the Ca^{2+} influx during a presynaptic action potential.

7. Suppose acetylcholine is released at two different synapses, synapse 1 and synapse 2. At synapse 1, acetylcholine directly binds to and opens Na^+ channels. At synapse 2, the postsynaptic acetylcholine receptors are G protein-coupled receptors that indirectly close K^+ channels.
   a. Synapse 1 produces an excitatory postsynaptic potential (epsp) and synapse 2 produces an inhibitory postsynaptic potential (ipsp).
   b. Both synapse 1 and synapse 2 produce ipsps.
   c. Both synapse 1 and synapse 2 produce epsps.
   d. Synapse 1 produces an ipsp, and synapse 2 produces an epsp.

8. Neurotransmitter release is triggered by
   a. Ca^{2+} influx through voltage-dependent Ca^{2+} channels that open during depolarization.
   b. Na^+ influx through voltage-dependent Na^+ channels that open during depolarization.
   c. binding of Ca^{2+} to molecules of neurotransmitter.
   d. Ca^{2+} influx into the synaptic cleft.

9. An ipsp could result from
   a. opening nonspecific cation channels that are equally permeable to Na^+ and K^+.
   b. opening either K^+ or Cl^- channels.
   c. opening channels that are selectively permeable to Na^+.
   d. neurotransmitter receptors that directly open an ion channel that is part of the receptor but not from neurotransmitter receptors that indirectly affect ion channels through second messengers.

10. In synaptic transmission
   a. neurotransmitter is released from the synaptic terminal by exocytosis.
   b. an excitatory postsynaptic potential promotes firing of action potentials by the postsynaptic cell, whereas an inhibitory postsynaptic potential makes it less likely that the postsynaptic cell will fire action potentials.
   c. neurotransmitter molecules released from the synaptic terminal are detected by receptor molecules in the membrane of the postsynaptic neuron.
   d. Ca^{2+} influx triggers fusion of synaptic vesicles with the membrane of the synaptic terminal.
   e. all of the above.

Short Answer

11. Identify the following: synaptic vesicle, synaptic cleft, neurotransmitter receptor.

12. What roles do voltage-dependent ion channels play in the neurotransmitter release?

13. List three ways of changing the postsynaptic ionic permeability to produce an excitatory postsynaptic potential.

14. List three substances that are used as neurotransmitter substances in the nervous system.

15. List three ways to prevent the occurrence of a postsynaptic potential at a synapse.

Essay

16. Describe the sequence of events between the arrival of an action potential at the synaptic terminal and the generation of the postsynaptic potential at a synapse.

Chapter Test Answers

1. **F**   2. **F**   3. **T**   4. **T**   5. **F**   6. **b**   7. **c**   8. **a**   9. **b**   10. **e**

11. Synaptic vesicle: the membrane-bound presynaptic structure in which neurotransmitter is stored; vesicles fuse with the plasma membrane to release neurotransmitter. Synaptic cleft: the extracellular space separating the presynaptic and postsynaptic cells at a synapse. Neurotransmitter receptor: a transmembrane protein in the postsynaptic cell at a synapse; includes a binding site for neurotransmitter and links transmitter binding to a change in postsynaptic permeability.

12. Voltage-dependent Na^+ channels are responsible for depolarization during the action potential and voltage-dependent Ca^{2+} channels respond to that depolarization by opening and allowing the Ca^{2+} influx that triggers release.

13. An excitatory postsynaptic potential could be produced by opening Na^+ channels, closing K^+ channels, or opening nonspecific cation channels.

14. A partial list of possible answers: acetylcholine, glutamate, glycine, GABA, dopamine, serotonin, norepinephrine.

15. Some possible answers: 1) Remove Ca^{2+} from the extracellular fluid; 2) apply a drug that blocks voltage-dependent Ca^{2+} channels of the synaptic terminal; 3) apply a drug that prevents synaptic vesicles from fusing with the plasma membrane; 4) apply a drug that blocks the neurotransmitter binding site of the postsynaptic receptor molecules; 5) apply a drug that blocks the ion channels that are affected by the neurotransmitter receptor; 6) in the case of indirect neurotransmitter actions, apply a drug that blocks the formation of the intracellular second messenger.

16. Depolarization produced by the action potential opens voltage-dependent Ca^{2+} channels in the plasma membrane of the synaptic terminal. Calcium ions flow into the terminal through the open Ca^{2+} channels and bind to sensor molecules associated with synaptic vesicles. Ca^{2+} binding triggers fusion of synaptic vesicles with the plasma membrane of the synaptic terminal, and the neurotransmitter within the vesicle is released into the synaptic cleft. Neurotransmitter binds to specific binding sites on the extracellular portions of receptor molecules in the plasma membrane of the postsynaptic cell. When the binding site is occupied by neurotransmitter, the receptor is activated and either directly or indirectly alters the ionic permeability of the postsynaptic cell. In directly neurotransmitter-gated channels, the ion channel is part of the receptor molecule and binding of the neurotransmitter opens the channel. In indirectly coupled receptors, the activated receptor sets in motion a series of intracellular signals, typically involving GTP-binding proteins and second messengers, which opens or closes ion channels in the postsynaptic cell. In either case, the change in ionic permeability alters the membrane potential and either promotes or discourages the firing of action potentials, depending on whether the membrane depolarizes or hyperpolarizes in response to the neurotransmitter. If the neurotransmitter increases p_{Na} or decreases p_K, an excitatory postsynaptic potential results. If the neurotransmitter increases p_K or p_{Cl} or decreases p_{Na}, then an inhibitory postsynaptic potential results.

Check Your Performance:

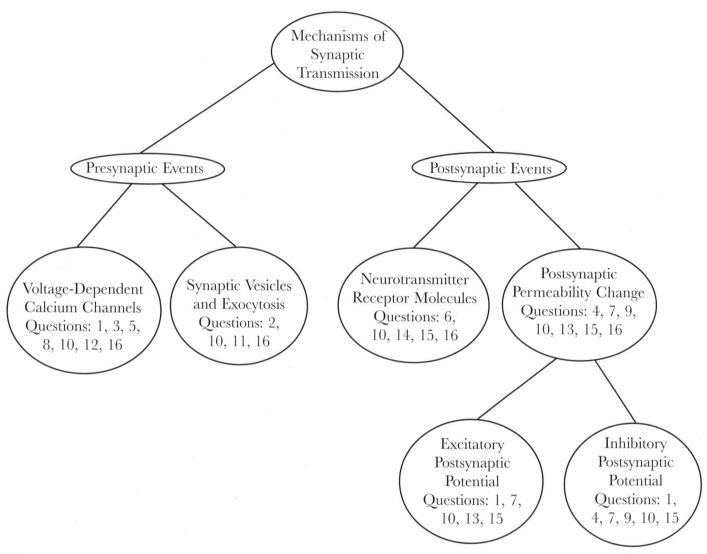

Note the number of questions in each grouping that you got wrong on the chapter test. Identify areas where your need further review and go back to relevant parts of this chapter.

Further help: If you continue to have difficulty with the concepts, a detailed presentation of these topics can be found in chapters 7 and 8 of *Neurobiology: Molecules, Cells, and Systems*, published by Blackwell Science, Inc.

UNIT III:
SENSORY
SYSTEMS

CHAPTER 6

General Characteristics of Sensory Systems

The sensory world of an organism is determined by the sensory cells with which it is endowed. If an organism is able to respond to a particular type of environmental stimulus, then it must possess sensory receptors that are sensitive to the physical energy of the stimulus. The perceived modality of a stimulus (i.e., whether it is perceived as light, touch, pain, etc.) is governed by the type of sensory neuron activated by the stimulus. Activity in neurons connected to photoreceptors is interpreted by the nervous system as providing information about light, activity in neurons connected to muscle spindle receptors is interpreted as giving information about muscle stretch, and so on.

ESSENTIAL BACKGROUND

- **The organization of the nervous system (Chapter 1)**
- **Mechanisms that control changes in membrane potential (Chapter 3)**
- **The action potential (Chapter 4)**
- **Synaptic transmission (Chapter 5)**

TOPIC 1: SENSORY RECEPTOR NEURONS

KEY POINTS

✓ *How do sensory receptor neurons translate a sensory stimulus into an electrical signal?*

✓ *How does the nervous system encode the intensity of a sensory stimulus?*

✓ *What is the receptive field of a sensory neuron?*

✓ *What classifications of sensory receptor neurons exist?*

Primary receptor neurons (also called **sensory receptors**) are the cells at the interface between the environment and the nervous system. Through the process of **sensory transduction**, these neurons translate mechanical, chemical, or light energy from the environment into an electrical signal that can be passed along in the nervous system. Sensory stimuli generate electrical signals called **receptor potentials** in the primary receptor neurons. As with other changes in membrane potential, the receptor potential is generated by opening or closing ion channels in the primary receptor neuron.

The receptor potential is graded according to stimulus intensity: Larger stimuli cause larger receptor potentials. The receptor potential in turn stimulates action potentials either in the primary sensory neuron itself or in higher order neurons that receive synaptic input from the primary receptor. The intensity of the stimulus is encoded by the frequency of action potentials. In sensory systems, some neurons increase their firing rate during a stimulus (the "on" pathway) and other neurons decrease their rate of firing during a stimulus (the "off" pathway). In both types, the more intense the stimulus, the larger the change in the rate of firing. During a sustained stimulus of constant intensity, the response of sensory neurons typically declines with time, a phenomenon called sensory adaptation. The **receptive field** of a sensory neuron is defined as the portion of the sensory surface (e.g., the skin, in the case of a touch-sensitive sensory neuron) where receipt of a stimulus affects the activity of the neuron.

Sensory receptors that monitor the external environment are called **exteroceptors**, and examples include receptors for vision, hearing, touch, and smell. Other receptor neurons, called **interoceptors**, monitor the internal environment of the organism, such as the blood pressure sensors of the carotid sinus. Another class is the **proprioceptors**, which monitor the position of the body in space or the position of joints and muscles.

Receptor neurons can also be classified based on the sensory stimulus to which they respond. **Mechanoreceptors** respond to mechanical displacement, including the muscle-spindle receptors that signal muscle stretch, touch and pressure receptors of the skin, and the hair cells found in the ear and the vestibular apparatus. The sensory cells of the visual system that translate light energy into electrical signals are called **photoreceptors**. **Chemoreceptors** are cells that detect and respond to chemical substances, whether they originate outside the body or inside the body. Sensory receptors that respond to tissue damage and give rise to the sensation of pain are called **nociceptors**, although they are probably chemoreceptors that respond to substances released by damaged cells. The skin also contains **thermoreceptors** that respond to skin temperature.

Topic Test 1: Sensory Receptor Neurons

True/False

1. The intensity of a sensory stimulus is encoded by the amplitude of the action potential generated by the stimulus in the primary receptor neuron.

2. Some sensory neurons decrease the rate of firing action potentials during a sensory stimulus.

3. Sensory adaptation refers to the fact that the nervous system associates activity in a particular sensory receptor neuron with a particular sensory sensation (e.g., activity in photoreceptors gives rise to the sensation of light rather than touch).

Multiple Choice

4. Sensory transduction
   a. occurs only in mechanoreceptor neurons but not in other types of sensory neurons.
   b. refers to the generation of postsynaptic potentials at the synapse between primary sensory neurons and higher order sensory neurons.
   c. is the translation of the energy of a sensory stimulus into an electrical signal in primary sensory neurons.
   d. all of the above.

5. The receptor potential
   a. is the change in membrane potential produced by a sensory stimulus in a primary sensory neuron.
   b. varies in amplitude with stimulus intensity.
   c. results from a change in ionic permeability produced by sensory transduction.
   d. all of the above.

Short Answer

6. Pressure-sensitive receptor neurons are examples of the class of receptors called _____. Primary sensory neurons that respond to painful stimuli are called _____.

7. Define the receptive field of a sensory neuron and give an example.

Topic Test 1: Answers

1. **False.** Action potentials have approximately constant amplitude. Thus, the intensity of a sensory stimulus is not encoded in the amplitude of the action potentials in sensory neurons. The nervous system uses a frequency code, in which the stimulus magnitude is encoded by the rate of action potentials. A weak stimulus produces action potentials at a low rate, and a stronger stimulus causes a higher frequency of firing.

2. **True.** Two types of response are observed in sensory neurons during a sensory stimulus: on responses and off responses. Neurons that exhibit off responses fire action potentials in the absence of a sensory stimulus, and the firing rate declines during a stimulus. Neurons that exhibit on responses fire at a low rate in the absence of a stimulus and increase the firing rate during a stimulus. In both cases, the intensity of the stimulus is coded by the degree of change (either decrease or increase) in the rate of firing.

3. **False.** Sensory adaptation refers to the declining response of sensory neurons during a sustained sensory stimulus. Because the frequency of action potentials in sensory neurons encodes the amplitude of the sensory stimulus, a decline in firing rate produces a corresponding reduction in perceived magnitude of sensation.

4. **c.** The nervous system uses electrical signals, which are produced by opening and closing ion channels. Primary sensory receptors generate an electrical signal in response to a particular type of sensory stimulus. The energy of the stimulus may be mechanical, chemical, or electromagnetic, and sensory transduction translates that form of energy into a change in ionic permeability and thus into an electrical signal.

5. **d.** In a primary receptor neuron, a sensory stimulus initiates sensory transduction, which culminates in opening or closing ion channels. The resulting change in ionic permeability produces the receptor potential, which is graded with the intensity of the sensory stimulus.

6. mechanoreceptors, nociceptors. Mechanoreceptors are sensory receptor cells that respond to mechanical displacement of the sensory surface. Painful stimuli activate nociceptors, which probably are chemoreceptors that detect chemicals released by damaged tissue.

7. The receptive field is the portion of the sensory surface where a stimulus affects the activity of the sensory neuron. For example, a touch-sensitive sensory neuron is stimulated

by displacement of the skin in the region of skin innervated by the sensory neuron terminate. Touch stimuli applied outside this region of skin will not affect the neuron.

TOPIC 2: PROCESSING SENSORY INFORMATION

KEY POINTS

✓ *How do synaptic connections alter the receptive field of higher order sensory neurons?*

✓ *What is lateral inhibition and what is its functional role in processing sensory information?*

The receptive field of a primary sensory neuron corresponds to the part of the sensory surface to which the neuron is directly connected. For instance, a touch-sensitive neuron responds to mechanical displacement of the skin at the location contacted by the neuron's processes in the skin. For higher order neurons, the organization of the receptive field is typically more complex, because it depends on the set of primary receptor neurons from which the higher order neuron receives direct or indirect synaptic connections.

A common synaptic interaction that alters responses of higher order sensory neurons is **lateral inhibition**, in which activity in sensory neurons receiving input from a particular region of the sensory surface inhibits sensory neurons receiving connections from adjacent parts of the sensory surface (**Figure 6.1**). Inhibition is mediated by inhibitory interneurons that extend to the side and make inhibitory synapses on neighboring pathways. Such lateral inhibitory connections are a common feature of sensory systems. Lateral inhibition alters the receptive fields of the higher order sensory neurons, producing **center-surround receptive fields** (see Figure 6.1). When a stimulus is placed in the center of the receptive field, it excites the second-order neuron, which receives direct excitatory synapses from the primary receptor cells in that region. When a stimulus is placed in the surrounding part of the receptive field, the second-order neuron is inhibited, because of the indirect inhibitory connection from the sensory cells that are excited by the stimulus in that region. Note that large stimuli covering both the center and surround regions are not very effective at exciting the second-order sensory neurons, even though the primary receptors are stimulated strongly by the stimulus. Lateral inhibition accentuates differential activity in an array of primary sensory receptors, enhancing the demarcation between stimulated and unstimulated regions of the sensory surface. **Contrast enhancement** of the border between stimuli of different intensities is an important feature of sensory systems.

Topic Test 2: Processing Sensory Information

True/False

1. The receptive field of a secondary sensory neuron is the same as the receptive fields of the primary sensory receptors from which the secondary neuron receives synapses.

2. Lateral inhibition enhances contrast between regions of the sensory surface receiving sensory stimuli of different strengths.

3. A center-surround receptive field arises because stimuli in certain parts of the receptive field excite primary receptor neurons, whereas stimuli in other parts of the receptive field inhibit primary receptor neurons.

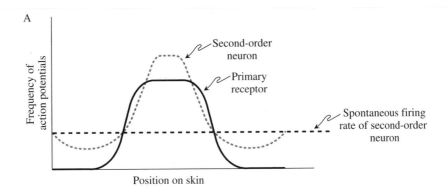

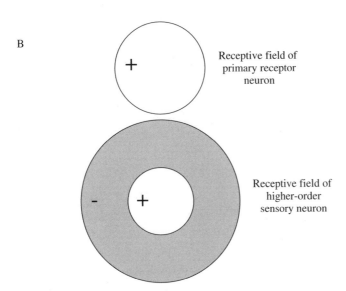

Figure 6.1 Receptive fields of primary and second-order sensory neurons are different because of lateral inhibition. A. The frequency of action potentials recorded in a primary sensory neuron (black line) and in the corresponding second-order neuron (gray line) in response to a stimulus applied to various positions on the skin. The dashed line shows the frequency of "spontaneous" action potentials in the second-order neuron in the absence of stimulation. B. View from above of the receptive fields of the primary sensory neuron (above) and the second-order neuron (below). The plus sign indicates that stimuli in that region excite the neuron, and a minus sign indicates that stimuli in that region inhibit the neuron.

Multiple Choice

4. A second-order sensory neuron has a center-surround receptive field, with an excitatory central region and inhibitory surround region. Which stimulus configuration would produce the highest firing rate?
   a. A stimulus covering the entire receptive field.
   b. A stimulus in the surround region, with no stimulus in the center.
   c. A stimulus in the central region, with no stimulus in the surround.
   d. Complete absence of a stimulus produces the highest firing rate, because all stimuli inhibit the cell.

5. Lateral inhibition
   a. reduces the ability of the nervous system to detect the exact location of a sensory stimulus, because it makes receptive fields larger.

b. refers to the inhibition of primary sensory neurons produced by stimuli located outside the receptive field of the neuron.

c. is a phenomenon associated only with the touch sensory system of the skin.

d. is mediated by means of inhibitory interneurons that are excited by stimuli at a particular location and make inhibitory synapses on sensory neurons of neighboring locations.

Short Essay

6. Briefly describe the synaptic connections underlying a center-surround receptive field (assume that the central region produces excitation).

Topic Test 2: Answers

1. **False.** The receptive field of a secondary sensory neuron can be quite different from the receptive fields of the primary receptor cells from which it receives input. The effect of a stimulus on higher order neurons depends on synaptic interactions and on the response properties of the primary receptor cells.

2. **True.** At the border between a strong and a weak stimulus, primary receptors on one side of the border are strongly activated, whereas those on the other side are weakly activated. Similarly, the second-order neurons that receive direct excitatory synapses from the primary receptor cells will be strongly excited on the one hand and weakly excited on the other. However, lateral inhibitory effects are asymmetrical at the border. Second-order neurons receiving the strong stimulus pass a large inhibitory signal to their neighbors, but second-order neurons receiving the weak stimulus pass a small inhibitory signal to their neighbors. Thus, on one side of the border second-order cells receive weak excitation and strong inhibition, whereas on the other side second-order cells receive strong excitation and weak inhibition. This produces a large contrast in the firing rate of the second-order cells at the point of transition from weak to strong stimulus, enhancing the ability of the nervous system to detect the change in stimulus intensity.

3. **False.** Center-surround receptive fields arise from synaptic interactions involving interneurons at stages of synaptic processing subsequent to the primary receptor neurons. Primary receptor neurons respond uniformly to sensory stimuli within their receptive fields.

4. **c.** A stimulus restricted to the excitatory central part of the receptive field, with no stimulus in the inhibitory surround region, would produce the highest firing rate. Such a stimulus activates primary receptor cells that make excitatory synapses on the second-order cell but do not activate the primary receptor cells that excite neighboring second-order neurons, which in turn inhibit the cell in question. A stimulus that covers the entire receptive field would be ineffective, because it activates both excitatory and inhibitory inputs to the cell. A stimulus restricted to the surround region inhibits the cell by activating only the lateral inhibitory inputs.

5. **d.** Lateral inhibition is a common feature of sensory systems, not just touch. As choice d correctly states, the inhibition is carried out by inhibitory interneurons whose connections spread laterally to neighboring sensory neurons.

6. The excitatory central region of a center-surround receptive field arises from excitatory synapses on the second-order cell from primary receptor neurons whose receptive fields coincide with the center of the center-surround receptive field. The inhibitory surround region arises from inhibitory synapses on the second-order sensory neuron made by lateral inhibitory neurons, which are in turn excited by other second-order neurons receiving excitatory synapses from primary receptors whose receptive fields are located in the surround of the center-surround receptive field.

IN THE CLINIC

Sensory systems provide the nervous system with information about the environment. This information arrives in the form of action potentials in the sensory neurons. The subjective sensation that results from those action potentials is determined by the identity of the sensory cell that is firing. In this **labeled line** method of sensory coding, action potentials in a touch-sensitive mechanoreceptor will give rise to sensation of touch in the receptive field of that cell, action potentials in a nociceptor will give rise to sensation of pain in the appropriate receptive field, and so on. This is true even if the receptive field for the sensory neuron no longer exists, as exemplified by the phenomenon of **phantom limb sensation** in amputees. When a limb is amputated, the nerves that innervate the limb are severed. Sometimes, however, the axons of the neurons in the cut nerve fire action potentials, because of local irritation of the cut end of the nerve. These action potentials are interpreted by the brain as sensations arising in the nonexistent limb. If action potentials arise in nociceptors of the missing limb, chronic pain is the result.

Chapter Test

True/False

1. In sensory neurons, intensity of a sensory stimulus is encoded by the frequency of action potentials, which may either increase or decrease with increasing intensity of the stimulus.

2. The modality of a sensory stimulus is encoded by the type of sensory neuron that fires in response to the stimulus.

3. Sensory transduction refers to propagation of the receptor potential along the axon of a primary sensory neuron.

4. The receptive field is the part of a primary receptor neuron that receives synaptic inputs from inhibitory interneurons.

5. Nociceptors respond to light touch applied to the skin.

6. Second-order sensory neurons commonly have different receptive fields from those of primary receptor neurons.

7. Lateral inhibition refers to the fact that activity in sensory neurons tends to inhibit the activity of adjacent sensory neurons.

Multiple Choice

8. The receptor potential
   a. is the change in membrane potential produced when a sensory stimulus opens or closes ion channels in a primary sensory neuron.
   b. is found in exteroceptors but not interoceptors or proprioceptors.
   c. is produced in second-order sensory neurons by inhibitory synapses.
   d. is found only in sensory neurons that have center-surround receptive fields.

9. The receptive field of a higher order sensory neuron
   a. consists of some regions where stimuli excite the neuron and other regions where stimuli inhibit the neuron.
   b. depends on the set of primary receptors from which the neuron receives synapses and on the organization of those synaptic connections.
   c. is often influenced by lateral inhibitory connections from interneurons and from direct synaptic connections from primary receptor neurons.
   d. all of the above.

10. Lateral inhibition
    a. is responsible for adaptation of primary sensory receptors during sustained sensory stimulation.
    b. is produced by inhibitory interneurons that receive excitatory synapses from a set of sensory neurons and make inhibitory synapses onto adjacent sensory neurons.
    c. prevents the occurrence of receptor potentials in primary sensory neurons.
    d. all of the above.

Short Answer

11. Identify the following: mechanoreceptor, nociceptor, chemoreceptor.

12. The process of translating stimulus energy into an electrical signal is called _____. This process occurs in _____.

13. Define the receptive field of a touch-sensitive sensory neuron of the skin.

Essay

14. Beginning with the receipt of the sensory stimulus by the primary receptors, describe the sequence of events underlying the generation of center and surround responses of a second-order sensory neuron that has a center-surround receptive field, with stimulation in the center producing excitation.

Chapter Test Answers

1. **T**   2. **T**   3. **F**   4. **F**   5. **F**   6. **T**   7. **T**   8. **a**   9. **d**   10. **b**

11. Mechanoreceptor: sensory receptor cell that responds to mechanical displacement. Nociceptor: sensory receptor cell that responds to painful stimuli. Chemoreceptor: sensory receptor cell that responds to chemical substances.

12. Sensory transduction, primary sensory receptors.

13. The receptive field of a touch-sensitive neuron of the skin is the region of the skin where application of a touch stimulus affects the firing of action potentials by the neuron. This corresponds to the area of the skin contacted by the peripheral processes of the neuron.

14. The center response originates from primary sensory neurons whose receptive fields coincide with the center. A stimulus produces a receptor potential in the primary sensory receptors. The receptor potential either triggers action potentials in the primary receptor or directly spreads to the synaptic terminal. The neurotransmitter released at this synapse depolarizes the second-order neuron and increases the rate of firing action potentials. The surround response begins in the same way, with a different set of primary sensory neurons whose receptive fields are located in the surround. These primary sensory receptors excite their own second-order neurons, which in turn activate inhibitory interneurons. The interneurons send processes laterally to make inhibitory synapses on surrounding second-order neurons.

Check Your Performance:

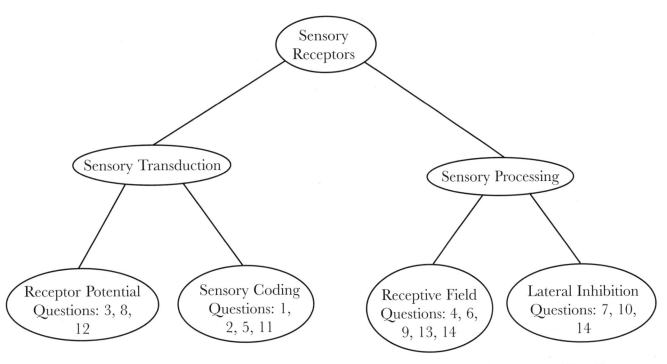

Note the number of questions in each grouping that you got wrong on the chapter test. Identify areas where you need further review and go back to relevant parts of this chapter.

Further help: If you continue to have difficulty with the concepts, a detailed presentation of these topics can be found in chapter 14 of *Neurobiology: Molecules, Cells, and Systems*, published by Blackwell Science, Inc.

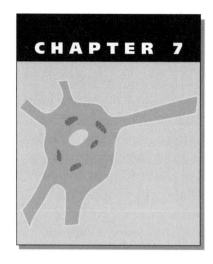

CHAPTER 7

The Somatic Senses

Collectively, the neural pathways concerned with the processing of signals from the receptors of the skin, muscles, and joints are called the **somatosensory** pathways. This chapter presents the different types of primary receptor involved in the somatosensory system and discusses the spinal cord and brain centers involved in processing somatosensory information.

ESSENTIAL BACKGROUND

- The anatomical organization of the spinal cord
- The anatomical organization of the brain, especially the thalamus and cortex
- General features of sensory systems (Chapter 6)
- Mechanisms of synaptic transmission (Chapter 5)

TOPIC 1: SOMATOSENSORY RECEPTOR NEURONS

KEY POINTS

✓ *What types of sensory receptor neurons are found in the skin?*

✓ *What kinds of sensory stimuli activate the skin receptors?*

✓ *What types of sensory receptor neurons innervate the muscles and joints?*

✓ *What sensory information is encoded by the various muscle and joint receptors?*

In mammals, five classes of skin mechanoreceptor are generally found: **Pacinian corpuscles**, **Meissner corpuscles**, **Ruffini corpuscles**, **Merkel receptors**, and **hair follicle receptors**. The hair follicle receptors spiral around single hair follicles and fire a burst of action potentials when the hair originating from that follicle is deflected. Pacinian corpuscles, Meissner corpuscles, and hair follicle receptors adapt rapidly during a sustained stimulus and so are particularly sensitive to vibrating touch stimuli. Ruffini corpuscles and Merkel receptors adapt slowly during sustained stimuli and thus encode intensity of maintained pressure. Axons of all skin mechanoreceptor neurons are moderately large and myelinated, conducting action potentials at a velocity of 30 to 70 meters per second (conduction velocity group II, also called group Aβ).

Thermoreceptors have numerous fine branches in the skin called free endings. **Cold receptors** are excited when the skin is cooled below body temperature and have myelinated axons that conduct more slowly than mechanoreceptors (conduction group III, also called group Aδ).

Warm receptors are excited when the skin is warmed above body temperature and have slowly conducting (<1 meter per second), unmyelinated axons of conduction group IV (also called C-fibers, or group C). **Nociceptors** also have free endings, which detect chemical substances released from damaged tissue and have either unmyelinated (group C) or small-diameter myelinated axons (group III). The faster-conducting nociceptors give rise to the flash of searing pain immediately after a wound, and the slow-conducting nociceptors produce persistent burning pain.

The **muscle spindle** is a fibrous capsule oriented in parallel with skeletal muscle fibers, containing specialized **intrafusal muscle fibers**. The nerve fibers of the muscle spindle receptor neurons form spiral endings (**annulospiral endings**) around the intrafusal muscle fibers. The axons of the muscle spindle receptors are the fastest conducting sensory axons of peripheral nerves (group Ia). The muscle spindle receptors provide information about muscle length. Because the length of a muscle can vary dramatically during use, the set point for activation of the muscle spindle receptor varies accordingly under control of a special group of motor neurons (the γ **motor neurons**), which innervate the intrafusal muscle fibers. **Golgi tendon organs** are encapsulated structures located at the junction between a skeletal muscle and its tendon. Tendon organ receptors provide information about the tension exerted by the muscle. The axons of the tendon organ receptor neurons are members of the second-fastest conducting group of sensory axons in peripheral nerves (group Ib). **Joint capsules** are located between bones at joints. Mechanoreceptor neurons of the joint capsule signal joint angle and have axons that conduct action potentials at a moderately fast speed (group II).

Topic Test 1: Somatosensory Receptor Neurons

True/False

1. All mechanoreceptors of the skin adapt rapidly during maintained pressure.

2. The somatosensory pathways carry sensory information from skin receptors but not from proprioceptors.

3. Muscle spindle receptors give information about muscle length, whereas Golgi tendon organ receptors give information about muscle tension.

4. The axons of all proprioceptors conduct at the same speed: All are members of conduction group Ia.

5. The five types of mechanoreceptor found in mammalian skin are Pacinian corpuscles, Meissner corpuscles, Ruffini corpuscles, Merkel receptors, and hair follicle receptors.

6. The axons of all muscle spindle receptors are in group Ia, the fastest conducting group of sensory axons.

Multiple Choice

7. Based on the type of sensory stimulus to which they respond, skin sensory neurons can be grouped into three classes:
   a. proprioceptors, muscle receptors, and tendon receptors.
   b. group Ia, group Ib, and group II receptors.

c. myelinated receptors, unmyelinated receptors, and C-fiber receptors.

d. mechanoreceptors, thermoreceptors, and nociceptors.

8. Two types of nociceptor neurons in the skin are distinguished based on

a. the strength of painful stimulus required to excite the nociceptor.

b. whether the sensory ending in the skin is a free ending or a corpuscle ending.

c. whether the nociceptor responds to a damaging level of heat applied to the skin.

d. none of the above.

Short Answer

9. Give two examples of sensory receptor neurons whose axons are unmyelinated (C-fibers).

10. You record action potential activity from a sensory neuron that innervates the skin. The firing rate increases dramatically when pressure is applied to the skin and firing rate declines only slightly during maintained pressure. What kind of sensory receptor neuron could this be?

Topic Test 1: Answers

1. **False.** Only hair follicle receptors, Meissner corpuscles, and Pacinian corpuscles adapt rapidly; Ruffini corpuscles and Merkel receptors adapt slowly.

2. **False.** The somatosensory pathways process information from all somatic senses. This includes information about proprioception (joint and muscle sensations) and information derived from skin receptors.

3. **True.** Muscle spindles are embedded within the muscle, but tendon organs are located at the junction between the muscle and its tendon. The muscle spindle is in parallel with the muscle fibers, and the Golgi tendon organ is in series with the muscle fibers. Therefore, the former senses the length of the muscle fibers, whereas the latter senses the tension created by contraction of the muscle.

4. **False.** Only muscle spindle receptors have axons of conduction class Ia. The axons of the tendon organ receptors conduct action potentials more slowly, in class Ib. The joint capsule receptors have axons that conduct still more slowly, in group II.

5. **True.** Mammalian skin has three types of encapsulated mechanoreceptors (Pacinian, Meissner, and Ruffini) and Merkel receptors and receptors of the hair follicles. The receptor types differ in their relative sensitivity to touch and pressure and in their degree of adaptation during maintained stimulation. Not all parts of the skin contain all five receptor types, and the density of the receptors also varies in different regions of the skin. The finger tips, for instance, have a high density of touch-sensitive receptors, each with a small receptive field, whereas the skin of the forearm has a low density of mechanoreceptors, each with a relatively large receptive field.

6. **False.** This answer may seem to contradict the information presented earlier. However, although it is true that muscle spindles are innervated by group Ia sensory fibers, muscle spindles of mammals receive an additional input from sensory fibers of group II. Group Ia fibers form annulospiral endings, called primary endings, on intrafusal muscle fibers,

and group II fibers form other spiral endings called secondary endings. The primary and secondary sensory neurons respond differently to stretch of the muscle. Primary neurons (group Ia) signal the change in length when the muscle is stretched, and secondary neurons (group II) signal static length of the muscle. Therefore, the primary sensory fiber is also called the dynamic fiber, and the secondary sensory fiber is called the static fiber.

7. **d.** Skin sensory receptors respond to mechanical stimuli (mechanoreceptors), to thermal stimuli (thermoreceptors), or to stimuli that damage the skin (nociceptors). The categories in statement a have to do with proprioception. Statements b and c refer to classifications of sensory neurons based on the speed of action potential conduction in their axons.

8. **d.** Statements a, b, and c are all incorrect. Two types of nociceptor are distinguished based on the conduction velocity of their axons. The nociceptors with faster conducting axons (group III or group Aδ) mediate withdrawal reflexes and produce rapid pain. The slower conducting nociceptors (group C) are responsible for the prolonged burning pain sensation.

9. Unmyelinated C-fibers carry action potentials for two types of sensory receptors: warm receptors of the skin and the slow-conducting class of nociceptors.

10. The sensory neuron is a pressure-sensitive mechanoreceptor that shows little adaptation during sustained stimulation. This could be either a Ruffini corpuscle receptor or a Merkel receptor.

TOPIC 2: SPINAL CORD AND BRAIN STEM PATHWAYS FOR SOMATOSENSORY INFORMATION

KEY POINTS

✓ *How are the somatosensory pathways organized in the spinal cord?*

✓ *What are the relay stations for somatosensory information in the brain stem?*

✓ *What is somatotopic organization?*

Figure 7.1 shows the organization of somatosensory pathways in the spinal cord. Cell bodies of somatosensory neurons are located just outside the spinal cord in the **dorsal root ganglia**. Axons of the sensory neurons enter the spinal cord via the dorsal roots and make synaptic connections that underlie spinal reflexes. In addition, sensory information is transmitted to the brain for further analysis. The ascending axons carrying information about touch, pressure, vibration, and proprioception to the brain are found mostly in the **dorsal columns** (Figure 7.1A). Most sensory axons in the dorsal columns are branches of primary sensory neurons, whose axons branch as they enter the spinal cord from the dorsal roots. One branch enters the spinal gray matter to make synaptic connections with interneurons and motor neurons, and the other branch enters the dorsal column on the same side of the spinal cord and ascends to the brain. In addition, the dorsal columns contain axons originating from spinal interneurons that receive synaptic inputs from the primary sensory neurons.

Another important group of ascending sensory axons is found in the **lateral sensory tract** (Figure 7.1B). Unlike the dorsal column, the axons in the lateral sensory tract are predominantly those of spinal interneurons, which receive synaptic inputs from pain- and temperature-sensitive primary receptors and to a lesser extent from tactile sensory neurons. The axons of the inter-

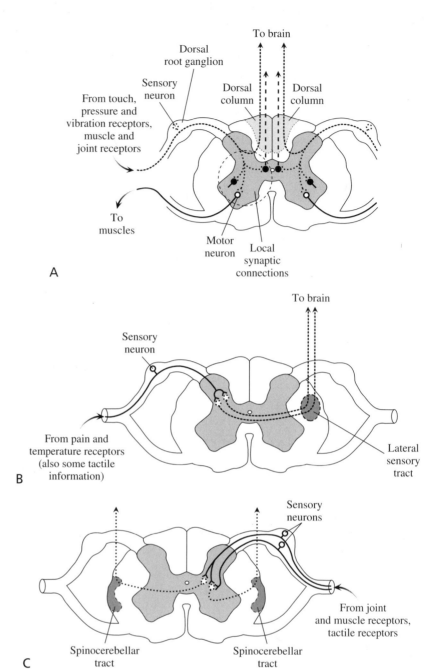

To brain

Dorsal
root ganglion

Sensory
neuron

From touch,
pressure and
vibration receptors,
muscle and
joint receptors

Dorsal
column

Dorsal
column

To
muscles

Motor
neuron

Local
synaptic
connections

A

To brain

Sensory
neuron

From pain and
temperature receptors
(also some tactile
information)

Lateral
sensory
tract

B

Sensory
neurons

From joint
and muscle receptors,
tactile receptors

Spinocerebellar
tract

Spinocerebellar
tract

C

Figure 7.1 A. The projection pattern of sensory information in the dorsal columns of the spinal cord. B. The projection pattern of sensory information in the lateral sensory tracts. C. The projection patterns of sensory information in the spinocerebellar tracts.

neurons cross the midline of the spinal cord before entering the lateral sensory tract on the opposite side of the spinal cord. A third path for ascending sensory axons is the **spinocerebellar tract** (Figure 7.1C), which is located in the lateral column on each side of the spinal cord. The axons in the spinocerebellar tracts come from spinal interneurons, some on the same side of the spinal cord and others from the contralateral side, which receive synapses from tactile receptors of the skin and from proprioceptors. As the name implies, the target of the spinocerebellar tract is the cerebellum.

Ascending sensory axons of the dorsal columns leave the spinal cord and terminate in the **dorsal column nuclei**, which are groups of interneurons near the dorsal surface of the medulla. Two dorsal column nuclei are found on each side of the medulla: The **gracile nucleus** is nearer to the midline, and the **cuneate nucleus** is located more laterally. Axons from the medial portion of the dorsal column on each side of the spinal cord target the gracile nucleus on the same side of the medulla, and axons from the lateral portions of the dorsal columns target the ipsilateral cuneate nucleus. Ascending the spinal cord from the sacral end, sensory axons are added successively at the lateral edge of the dorsal columns at each vertebral segment. Thus, at the cervical end of the spinal cord, sensory axons from the sacral spinal segments are located most medially, followed sequentially by axons from the lumbar, thoracic, and cervical segments moving laterally in the dorsal column. Because the sensory neurons of the lower limbs, lower trunk, upper trunk and limbs, and neck and face enter the spinal cord at successively higher spinal segments, the medial-lateral organization of sensory axons in the dorsal columns translates into a spatial map of the body surface within each dorsal column, with lower body parts represented medially and upper body parts represented laterally. This somatotopic map of the body within the nervous system is carried over into the synaptic projection to the dorsal column nuclei of the medulla: The lower portion of the body is represented in the gracile nucleus and the upper portion in the cuneate nucleus.

Many axons of the lateral sensory tract terminate in the **reticular formation** of the medulla, pons, and midbrain. The reticular formation sends widespread connections to the midbrain and forebrain, controlling the arousal of the animal. Painful stimuli arouse the animal by means of these reticular formation connections. Neurons of the reticular formation also relay pain and temperature information to thalamocortical pathways of sensory perception. In mammals, some ascending axons of the lateral sensory tract bypass the brain stem and pass directly into the thalamocortical system. This direct projection includes axons of spinal interneurons that receive inputs from the fast-conducting nociceptors (Aδ fibers) and is responsible for rapid perception of pain.

Topic Test 2: Spinal Cord and Brain Stem Pathways for Somatosensory Information

True/False

1. A touch-sensitive primary sensory neuron can have a very long axon, extending from its receptive field in the skin, into the spinal cord, and up to the brain.

2. Axons in the dorsal columns are exclusively those of skin receptors that carry information about touch, pressure, and vibration.

3. The main spinal cord pathway for ascending pain and temperature information is the spinocerebellar tract.

4. In the spinal cord, primary sensory neurons that are sensitive to muscle stretch and primary sensory neurons that are sensitive to tactile stimulation of the skin never make synapses onto the same interneuron.

5. The somatotopic map is the portion of the body innervated by the sensory neurons of a particular spinal segment.

6. The gracile nucleus and the cuneate nucleus are targets of sensory axons ascending in the lateral sensory tract.

7. All nociceptive information passes through neurons of the reticular formation before being relayed to other sensory centers of the brain.

Multiple Choice

8. Which of the following spinal cord pathways is most likely to contain axons of primary sensory neurons?
   a. lateral sensory tract
   b. spinocerebellar tract
   c. dorsal column
   d. ventral root

9. Which of the following spinal cord pathways contains axons of spinal interneurons, which in turn receive inputs from primary sensory neurons?
   a. lateral sensory tract
   b. spinocerebellar tract
   c. dorsal column
   d. all of the above

Short Essay

10. In the dorsal column somatosensory pathway, at what level of the pathway would you expect that lateral inhibition and center-surround receptive fields would first be observed?

Topic Test 2: Answers

1. **True.** The cell body of a touch-sensitive primary sensory neuron is located in the dorsal root ganglion. The axon exits the ganglion and extends a branch through the appropriate peripheral nerve to the receptive field in the skin. Another axon branch enters the spinal cord through the dorsal root. In the spinal gray matter, the axon makes numerous branches that synapse with interneurons and motor neurons. In addition, an axon branch of most tactile sensory neurons enters the dorsal column and ascends the length of the spinal cord to the dorsal column nuclei of the medulla. Thus, primary sensory neurons can have exceedingly axons that extend from the skin to the brain stem.

2. **False.** The dorsal columns are also the projection path for sensory information originating from muscle and joint receptors. Therefore, both proprioception and tactile senses are represented in the dorsal column pathway. Also, axons of spinal interneurons are also found in the dorsal columns, and so the columns do not exclusively contain axons of primary sensory neurons.

3. **False.** Pain and temperature information is transmitted to the brain by the axons of the lateral sensory tract.

4. **False.** In many cases, various sensory modalities are in fact strictly segregated as the statement specifies. However, in other instances, spinal interneurons receive synaptic inputs from more than one type of somatosensory receptor. Such interneurons are called multimodal interneurons, and they are especially common among interneurons whose axons make up the spinocerebellar tract. Multimodal interneurons provide information about a confluence of sensory events, such as contact of a limb with the ground during locomotion (tactile information) combined with proprioceptive information about muscles of the same limb.

5. **False.** A somatotopic map refers to the systematic mapping of the body onto the anatomical structure of the nervous system. In the spinal cord, a simple example is the spatial segregation of sensory inputs from various body parts into different portions of the dorsal columns in the cervical spinal cord. The lower parts of the body are represented medially in the dorsal column, and the upper parts of the body are represented laterally. This somatotopic map is also carried over into the dorsal column nuclei that receive synaptic inputs from the dorsal columns.

6. **False**. The gracile nuclei and the cuneate nuclei are the dorsal column nuclei of the medulla. As the name implies, these nuclei are the targets for the axons of the dorsal columns, not the axons of the lateral sensory tract.

7. **False.** The reticular formation is in fact a major target for sensory axons carrying information about pain in the lateral sensory tract. However, the fast-conducting nociceptors make synaptic connections onto spinal interneurons whose axons pass through the brain stem and project directly to the thalamocortical system. Therefore, some pain information is not relayed through the reticular formation.

8. **c.** Ascending axons of primary sensory neurons make up most of the dorsal columns. The lateral sensory tract and the spinocerebellar tract consist of axons of spinal interneurons, not axons of primary sensory neurons. The ventral root of each spinal segment carries outgoing axons of motor neurons and is not a sensory structure.

9. **d.** Although the dorsal column contains mostly axons of primary sensory neurons, some axons in the dorsal columns are those of interneurons. The lateral sensory tract and the spinocerebellar tract consist of the axons of sensory interneurons.

10. Inhibitory synaptic interactions are required for the construction of center-surround receptive fields, and the first opportunity for such interactions is, for the most part, at the level of the dorsal column nuclei, where the first synapses are made by the primary sensory neurons in dorsal column pathway. However, axons of spinal interneurons are also found in the dorsal columns, and for these cells, lateral interactions are possible in the spinal cord before the outgoing axons enter the dorsal columns.

TOPIC 3: THALAMOCORTICAL PROCESSING OF SOMATOSENSORY INFORMATION

KEY POINTS

✓ *How is somatosensory information sent from the brain stem to the cerebral cortex?*

✓ *What part of the thalamus is involved in processing somatosensory information?*

✓ *What part of the cerebral cortex is involved in processing somatosensory information?*

✓ *What is the somatotopic organization of the sensory cortex?*

The neurons of the gracile and cuneate nuclei have axons that cross the midline of the medulla (i.e., the axons decussate) and form a large bundle on each side of the brain called the **medial lemniscus**. The axons of the medial lemniscus move progressively dorsally and laterally as they ascend through the midbrain and ultimately make synapses on relay neurons of the **ventral posterior thalamus**. The thalamus is a complex group of nuclei in the diencephalon that process and relay various kinds of sensory information. Because of the decussation of axons in the medial lemniscus, the thalamus on one side of the body receives tactile and proprioceptive information from the opposite side of the body. The lateral-medial somatotopic map observed in the dorsal column nuclei is also evident in the ventral posterior thalamus, but the direction of mapping is reversed. Tactile information from the hindlimbs projects to the most lateral portion of the thalamus, followed by information from the trunk and then the forelimbs at more medial positions. In addition, the medial part of the ventral posterior thalamus receives tactile inputs from the face, which enter the brain via the trigeminal nerve (cranial nerve V).

From the ventral posterior thalamus, somatosensory information is relayed to the cerebral cortex. The main cortical target is the **postcentral gyrus** (**Figure 7.2**), which is an outfolding (gyrus) located just behind the deep groove (the central sulcus) at the approximate anterior-posterior midpoint of the brain. The postcentral gyrus is the primary somatosensory area of the cortex. Because the thalamus on each side projects ipsilaterally to the cortex, the sensory cortex on the right side of the brain receives sensory information from the left side of the body and vice versa.

Just as in the dorsal column nuclei and thalamus, a somatotopic map exists in the primary somatosensory cortex. The face, lips, and tongue are represented in lateral portions of the primary somatosensory cortex, followed progressively by the upper limb, trunk, and lower limb at more medial positions (see Figure 7.2). The amount of somatosensory cortex devoted to a particular part of the body's surface does not correspond to the relative size of the body part. Instead, it corresponds to the number of sensory neurons that innervate that region. Thus, the fingers and tongue receive a large representation in the human cortex, whereas the trunk and legs occupy a relatively small portion of the cortex.

In the tactile sensory system of the skin, lateral inhibitory interactions occur at all synaptic relay stations: the dorsal column nuclei, thalamus, and cortex. Because of lateral inhibition, many cortical somatosensory neurons have center-surround receptive fields. Some cortical neurons have more complicated receptive fields. For instance, some cortical neurons respond best to a stimulus moving along the skin in a particular direction, whereas a stimulus moving in the opposite direction or a stationary stimulus is largely ineffective.

Topic Test 3: Thalamocortical Processing of Somatosensory Information

True/False

1. Somatosensory information reaching the left thalamus originates from the right side of the body.

2. Neurons of the dorsal column nuclei send axons directly to the primary somatosensory cortex.

3. The posterior portion of the ventral thalamus is a relay station for tactile and proprioceptive sensory information.

4. A somatotopic map of the body surface is found in the primary sensory cortex but not in the thalamus or the dorsal column nuclei.

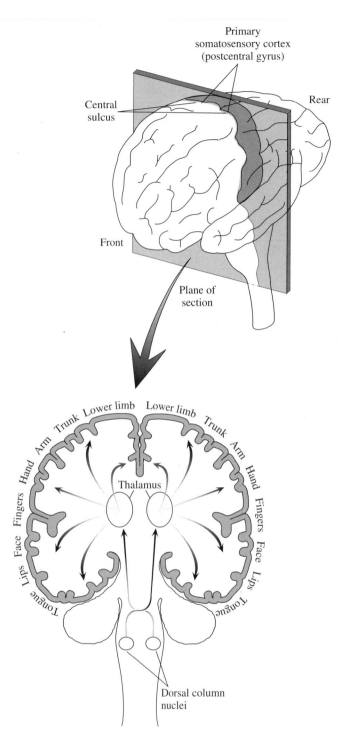

Figure 7.2 The somatotopic organization of somatosensory information in the primary somatosensory cortex, which is located just posterior to the central sulcus (upper diagram). The projections from the various parts of the body onto the somatosensory cortex are shown in the lower diagram, which depicts a section through the brain in the plane illustrated in the upper diagram. Because of the decussation of the ascending axons from the dorsal column nuclei in the brain stem, the sensory information on one side of the cortex comes from the contralateral side of the body.

5. The primary somatosensory cortex is found in the prefrontal gyrus, just in front of the central sulcus.

6. Body parts are represented in the somatosensory cortex in proportion to their size.

Multiple Choice

7. The thalamocortical pathway
   a. carries sensory information from the cortex to the spinal cord.

b. is the target of sensory information from the dorsal columns and from the lateral sensory tract.

c. does not transmit proprioceptive information.

d. all of the above.

8. The primary somatosensory cortex
a. has a somatotopic representation of the body.
b. has neurons with complex receptive fields.
c. is located in the postcentral gyrus of the cerebral cortex.
d. all of the above.

Short Answer

9. The axon pathway leading from the dorsal column nuclei to the thalamus is the _____ (two words).

10. The part of the thalamus that receives tactile sensory input from the dorsal column pathway is the _____ (three words).

Topic Test 3: Answers

1. **True.** Crossover of sensory information from one side of the body to the opposite side of the brain is a common feature of somatosensory pathways. Before sensory signals reach the thalamus, the crossover is complete for all somatosensory modalities.

2. **False.** Axons from the dorsal column nuclei project to the thalamus and terminate there. Thalamic neurons then relay the sensory signals to the cortex.

3. **True.** The ventral posterior nucleus of the thalamus is the target for axons of the medial lemniscus, which transmit proprioceptive and tactile information from the dorsal column nuclei.

4. **False.** Somatotopic maps are found not only in the cortex but also in the thalamus and dorsal column nuclei. In all cases, there is a medial-lateral representation of the body surface, although the orientation of the map is reversed in the thalamus (hindlimb = lateral) compared with the dorsal column nuclei and primary somatosensory cortex (hindlimb = medial).

5. **False.** The primary somatosensory cortex is actually located just behind the central sulcus in the postcentral gyrus.

6. **False.** Although a systematic representation of the body exists in primary somatosensory cortex, the amount of space devoted to a particular body part is not determined by the absolute size of the part. Instead, the representation in the cortex depends on the number of sensory neurons devoted to the body part.

7. **b.** The thalamus is the target of projections from the dorsal column nuclei and from lateral tract axons of interneurons that receive inputs from the fast-conducting nociceptors. In statement a, the direction of information flow is reversed. Statement c is false because the dorsal column projection to the thalamus includes proprioceptive information.

8. **d.** Primary somatosensory cortex is located in the postcentral gyrus of the cerebral cortex. A somatotopic representation of the body extends from the midline to the most lateral portions of the primary somatosensory cortex, and neurons in the primary somatosensory cortex can have complex receptive fields, such as directionally selective movement sensitivity.

9. **medial lemniscus**. This is the name of the large bundle of axons that extends from the dorsal column nuclei, through the midbrain, and into the diencephalon.

10. **ventral posterior thalamus**. This is the portion of the thalamus that receives the axons of the medial lemniscus, which originate from the dorsal column nuclei.

IN THE CLINIC

Spatial segregation of various sensory pathways in the spinal cord can provide important clues in diagnosis of spinal cord disease or damage. For example, tingling and/or numbness in the arms and legs on both sides of the body would suggest a possible problem in the dorsal part of the upper spinal cord, interfering with the transmission of signals in the dorsal columns on both sides of the cord. If the symptoms were restricted to one side of the body, it would suggest localization of the lesion to the same side of the spinal cord, because the dorsal column projections are ipsilateral within the cord. If damage were in the lateral part of the spinal cord instead of the dorsal part, disturbance in pain and temperature sensation on the contralateral side of the body might result if transmission in the lateral sensory tract is disrupted. Combined with a more complete neurological examination to test for possible reflex and other motor disturbances, careful analysis of sensory deficits can help guide more direct diagnostic tools such as magnetic resonance imaging.

Chapter Test

True/False

1. Most axons in the dorsal columns of the spinal cord are branches of axons of primary sensory neurons.

2. The cell bodies of the primary sensory neurons are located in the gray matter of the spinal cord.

3. The medial lemniscus is an axon tract within the spinal cord carrying information from skin mechanoreceptors.

4. The lateral sensory tract of the spinal cord carries ascending sensory information about pain and skin temperature.

5. Action potentials of Pacinian corpuscle receptors originate in the peripheral ending in the skin and propagate directly into the brain, without passing through an intervening synaptic relay.

6. All of the different types of skin mechanoreceptors adapt slowly during sustained stimulation and have small unmyelinated axons of the slow-conducting group C.

7. In mammals, all pain sensory information is relayed directly from the spinal cord to the thalamus, without passing through the reticular formation of the brain.

8. Sensory information carried in the lateral sensory tract on the left side of the spinal cord originates on the right side of the body, whereas sensory information carried in the dorsal column on the left side of the spinal cord originates on the left side of the body.

9. The axons of primary sensory neurons entering the spinal cord either terminate within the spinal cord or project into the brain, but not both.

10. Free nerve ending in the skin are associated with thermoreceptors and nociceptors.

Multiple Choice

11. Which of the following sensory receptors is involved in proprioception?
    a. Meissner corpuscle
    b. Golgi tendon organ
    c. muscle spindle receptor
    d. a and c, but not b
    e. b and c, but not a

12. Lateral inhibition and center-surround receptive fields could be observed at which of the following levels?
    a. dorsal column nuclei
    b. thalamus
    c. primary somatosensory cortex
    d. all of the above
    e. none of the above

13. In the primary somatosensory cortex, sensory representation of the face is found in the
    a. part closest to the midline.
    b. most lateral part of the cortex.
    c. in the middle part of the cortex.
    d. all of the above
    e. there is no sensory representation of the face in the somatosensory cortex.

14. Tactile information from the skin is carried in which of the following ascending spinal cord tracts?
    a. dorsal column
    b. spinocerebellar tract
    c. lateral sensory tract
    d. all of the above
    e. none of the above

Short Answer

15. The principal ascending pathway for skin mechanoreceptor information in the spinal cord is the _____ (two words). The ascending axons from this pathway terminate in the _____ (three words), located in the _____ (one word; part of the brain).

16. Identify the following: cuneate nucleus, postcentral gyrus, spinocerebellar tract.

Essay

17. Briefly trace the ascending pathways for pain, starting with the primary nociceptors and ending in the cerebral cortex.

18. What anatomical arrangement accounts for the fact that touch-sensitive neurons of the thalamus respond to stimuli applied to the contralateral side of the body?

19. Why do you think that it is functionally important for muscle spindle receptors to have axons of group Ia instead of group C?

20. Briefly describe the ascending pathway for somatosensory information from skin mechanoreceptors to the cerebral cortex.

Chapter Test Answers

1. **T**   2. **F**   3. **F**   4. **T**   5. **T**   6. **F**   7. **F**   8. **T**   9. **F**   10. **T**   11. **e**   12. **d**

13. **b**   14. **d**

15. dorsal column, dorsal column nuclei, medulla

16. Cuneate nucleus: one of dorsal column nuclei (the other being the gracile nucleus); it is the more lateral of the nuclei and receives synaptic input from the axons of the lateral part of the dorsal column. Postcentral gyrus: the cortical gyrus located just behind the central sulcus; location of the primary somatosensory cortex. Spinocerebellar tract: located at the lateral edge of the spinal cord; carries ascending axons of spinal interneurons to the cerebellum of the brain stem; a variety of sensory modalities are represented, including multimodal interneurons.

17. Cell bodies of nociceptors are located in the dorsal root ganglia and have axons that terminate as free endings in the skin on one end and within the gray matter of the spinal cord on the other end. Spinal interneurons that receive synaptic inputs from nociceptors have axons that cross the midline and ascend in the lateral sensory tract of the contralateral spinal cord. The ascending axons terminate in the reticular formation or, in the case of fast pain receptors in mammals, bypass the reticular formation and project directly to the thalamus. Pain information from the reticular formation is also relayed to the thalamus. Thalamic neurons in turn send axons to the cerebral cortex.

18. Most touch information reaches the thalamus by means of projections of the dorsal column pathway. In this pathway, axons of sensory neurons ascend in the ipsilateral spinal cord and synapse in dorsal column nuclei of the ipsilateral medulla. However, axons of neurons in the dorsal column nuclei project to the thalamus through the medial lemniscus, which crosses to the opposite of the brain (decussates) in the brain stem. Because of this decussation, the thalamus on each side of the brain receives information from the contralateral body. Some touch information ascends in the lateral sensory tract as well. In this case, decussation of axons occurs in the spinal cord because spinal interneurons send their axons into the contralateral lateral sensory tract. In both cases, the decussation occurs at the level of the first set of interneurons after the primary receptors.

19. Group Ia sensory axons are the fastest conducting sensory axons of peripheral nerves, whereas C-fibers are the slowest conducting sensory axons. Muscle spindle receptors

report stretch of a muscle, such as occurs when a load is suddenly applied to a joint. The receptors elicit a compensating contraction of the muscle to maintain "correct" muscle length against the increased load. The distance from the muscle may be quite long, and C-fibers conduct action potentials at less than 1 meter per second, and so an action potential would take a second or more to reach the spinal cord after stretch of the muscle. This is too long for the information to be useful for regulation of muscle length. By contrast, an action potential in an axon of group Ia would reach the spinal cord in 10 milliseconds or less. This high conduction speed allows the regulation of muscle length to occur on a rapid time scale.

20. Cell bodies of skin mechanoreceptors are located in the dorsal root ganglia, and their central axon branches enter the spinal cord via the dorsal roots. In the spinal cord, the axons branch and send a branch into the dorsal column on the same side of the spinal cord. The ascending branches terminate in the dorsal column nuclei of the lower medulla, either in the gracile nucleus (sacral and lumbar spinal segments) or in the cuneate nucleus (thoracic and cervical spinal segments). The axons from the dorsal column nuclei form the medial lemniscus, which crosses the midline and projects to the contralateral ventral posterior thalamus in the diencephalon. The thalamic relay neurons then send axons to the ipsilateral primary somatosensory cortex (postcentral gyrus). In addition, interneurons that receive inputs from primary mechanoreceptor neurons within the spinal cord have axons that cross the midline and ascend to the brain within the lateral sensory tract.

Check Your Performance:

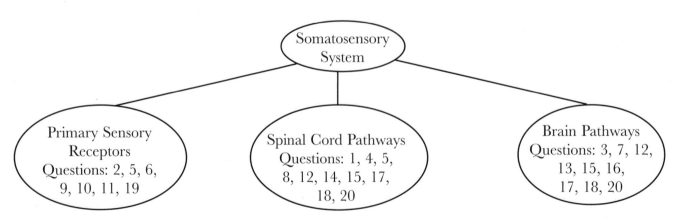

Note the number of questions in each grouping that you got wrong on the chapter test. Identify areas where you need further review and go back to relevant parts of this chapter.

Further help: If you continue to have difficulty with the concepts, a detailed presentation of these topics can be found in chapter 15 of *Neurobiology: Molecules, Cells, and Systems*, published by Blackwell Science, Inc.

CHAPTER 8

The Visual System

This chapter discusses the organization of the mammalian visual system, starting with the retina, which translates light energy into an electrical signal (phototransduction) and begins the task of extracting information about the outside world from those electrical signals. From the retina, visual signals are passed on to the thalamic and cortical centers that carry out additional analysis of the sensory information.

ESSENTIAL BACKGROUND

- **The anatomy of the eye**
- **The anatomical organization of the brain, especially the thalamus and cortex**
- **G proteins (GTP-binding proteins) and receptors coupled to G proteins**
- **Ionic permeability and membrane potential**

TOPIC 1: PHOTOTRANSDUCTION

KEY POINTS

✓ *What is the structure of photoreceptor cells?*

✓ *How is light absorbed by the photoreceptor cells?*

✓ *How is absorption of light coupled to a change in membrane potential of the photoreceptor?*

Translation of light energy into electrical signals is called **phototransduction** and is carried out by photoreceptors located in the retina at the back of the vertebrate eye. The two classes of photoreceptor are **rods**, which function in dim light, and **cones**, which function in brighter light. Photons of light are absorbed by visual pigment molecules in the photoreceptors. The pigment molecule of the rods is **rhodopsin**, which consists of a covalent linkage of the light-absorbing chromophore, **retinal** (vitamin A aldehyde), and a membrane protein, **opsin**.

When the chromophore absorbs a photon, it undergoes isomerization from a folded form (11-cis retinal) to a straight form (all-trans retinal). This photoisomerization triggers a conformation change in opsin, which is then able to activate a GTP-binding protein called **transducin**. Transducin then activates phosphodiesterase, an enzyme that hydrolyzes cyclic GMP to GMP. Thus, absorption of light decreases the intracellular concentration of cyclic GMP in the photoreceptor. Cyclic GMP is an internal messenger that opens nonspecific cation channels in the photoreceptor. Therefore, in darkness—when cyclic GMP levels are high—the channels are open and the photoreceptor is depolarized; in light—when cyclic GMP levels are low—the channels are closed and the photoreceptor membrane potential is more negative.

Topic Test 1: Phototransduction

True/False

1. The membrane potential of photoreceptors is more positive in darkness than in light.

2. The ion channels that underlie the light response of photoreceptors are Na^+ channels that allow only Na^+ to cross the plasma membrane.

3. Light is absorbed in the photoreceptor by the opsin protein molecule.

4. Rhodopsin is the visual pigment molecule found in rod photoreceptors.

5. Photoisomerization refers to the change in conformation of retinal from the all-trans isomer (straight form) to the 11-cis isomer (folded form) when it absorbs a photon of light.

Multiple Choice

6. Upon absorption of light by rhodopsin, the next step in phototransduction is
   a. the opsin protein undergoes a conformation change and activates the G protein, transducin.
   b. the opsin protein undergoes a conformation change, and the nonspecific cation channel formed by opsin closes.
   c. the opsin protein dissociates from the chromophore, and opsin then interacts with and activates phosphodiesterase.
   d. photoactivated rhodopsin acts as an enzyme that reduces the concentration of cyclic GMP.

7. The ion channels of the outer segment that close in response to illumination are opened by
   a. direct interaction of rhodopsin with the channel, which is prevented when rhodopsin undergoes photoisomerization.
   b. direct binding of the G protein, transducin, to the channel.
   c. direct binding of the intracellular messenger, cyclic GMP, to the channel.
   d. all of the above.

Essay

8. Phototransduction has several similarities to synaptic transmission. Discuss this statement, indicating specific similarities and dissimilarities.

Topic Test 1: Answers

1. **True.** Nonspecific cation channels in the plasma membrane of the outer segment are open in darkness, keeping the cell depolarized. Illumination closes the channels, removing their depolarizing influence.

2. **False.** The ion channels closed by light are actually nonspecific cation channels. This is similar to excitatory synaptic transmission, where opening nonspecific cation channels depolarizes the postsynaptic cell (Chapter 5).

3. **False.** Although opsin is a vital part of the pigment molecule, the photon is actually absorbed by the chromophore group that is covalently linked to opsin. The chromophore is a simple organic molecule, retinal (an abbreviation for retinaldehyde), which is derived from vitamin A.

4. **True.** Separate forms of opsin are found in rod and cone photoreceptors. When combined with retinal, opsin of rods forms the visual pigment rhodopsin, which absorbs light best in the blue-green part of the visual spectrum. In the human retina, there are three other types of opsin, each found in a separate type of cone. When combined with retinal, each of the cone opsins forms a different visual pigment molecule with a distinct wavelength specificity. The three cone pigments form the basis for human color vision.

5. **False.** Retinal undergoes a conformation shift upon absorbing a photon. However, the photoisomerization involves a shift from the 11-cis to the all-trans isomer.

6. **a.** Photoisomerization of retinal induces a conformation change in opsin, which is then able to interact with and activate the G protein transducin. Activated transducin then activates phosphodiesterase, an enzyme that reduces the concentration of cyclic GMP.

7. **c.** Cyclic GMP directly binds to protein making up the ion channel. Like neurotransmitter-gated channels, the channel opens when the binding site is occupied. The level of cyclic GMP in the outer segment is higher in darkness than in light. Thus, the binding site is more likely to be occupied by cyclic GMP in darkness, and more channels are open. Although both rhodopsin and transducin participate in the phototransduction process, neither interacts directly with the ion channels.

8. There are several similarities between synaptic transmission and phototransduction. In synaptic transmission, neurotransmitter molecules can affect ion channels either directly, by binding to and opening the channel, or indirectly, by activating a receptor that is coupled to G proteins. In phototransduction, both types of mechanism are observed. First, rhodopsin is coupled to a G protein (transducin), which in turn mediates the intracellular actions of photoactivated rhodopsin. Opsin is in fact structurally related to the neurotransmitter receptor molecules that mediate indirect actions of neurotransmitters at synapses. We can view the retinal chromophore group as being the "neurotransmitter" that activates the "receptor" (the opsin protein). In phototransduction, however, the "neurotransmitter" is covalently coupled to the binding site. In darkness, retinal is in an inactive form (the 11-cis isomer) that cannot activate opsin even though it occupies the binding site. In this view, light can be thought of as converting the "neurotransmitter" into its active all-trans form. Second, the light-sensitive ion channels of the outer segment are analogous to ion channels that are directly gated by binding of neurotransmitter. In the case of phototransduction, the "neurotransmitter" is cyclic GMP, and the binding site is located on the intracellular surface of the channel rather than on the extracellular face as in neurotransmitter-gated channels. As with many channels that mediate excitatory neurotransmission, the channel gated by cyclic GMP is a nonspecific cation channel.

TOPIC 2: SYNAPTIC ORGANIZATION OF THE RETINA

KEY POINTS

✓ *What types of neurons are found in the retina?*

✓ *What are the synaptic interactions among the different classes of retinal neurons?*

✓ *What are the receptive fields of the various retinal neurons?*

✓ *How is color information coded in the retina?*

Figure 8.1 shows the neurons of the retina. Photoreceptors release glutamate onto **bipolar cells** and **horizontal cells**. Glutamate is released in darkness, when the photoreceptors are depolarized. Horizontal cells are depolarized by glutamate, which opens nonspecific cation channels, and so horizontal cells hyperpolarize in response to illumination. The postsynaptic response of bipolar neurons is of two types. Like horizontal cells, **off-type bipolar neurons** have glutamate-gated nonspecific cation channels, are depolarized in darkness, and hyperpolarize in response to light. **On-type bipolar cells** depolarize in response to illumination as cyclic GMP-gated channels open. This occurs because glutamate reduces cyclic GMP concentration in the bipolar neurons via a GTP-binding protein and phosphodiesterase in a manner analogous to phototransduction.

Horizontal cells are lateral inhibitory interneurons, which release the neurotransmitter GABA. The receptive field of the bipolar neuron has a **center-surround organization**, with a central region where light either depolarizes the cell (on-type bipolar cells) or hyperpolarizes the cell (off-type bipolar cells) and a larger concentric region where light produces the opposite

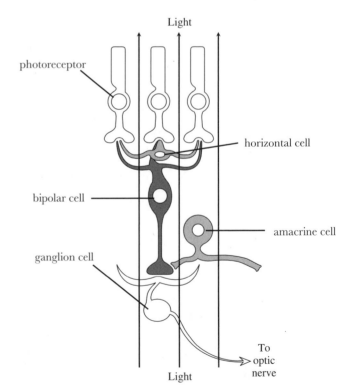

Figure 8.1 The neurons of the retina. The photoreceptors are at the rear of the retina and make synaptic connections with horizontal cells and bipolar cells. The bipolar cells then connect with the amacrine cells and ganglion cells in the inner portion of the retina. The axons of the ganglion cells make up the optic nerve.

effect. This opposing effect from the surrounding portion of the receptive field is mediated via the lateral synapses made by the horizontal cells.

Bipolar cells carry light signals to third-order neurons in the retina, the **amacrine cells** and the **ganglion cells**. Amacrine cells are lateral inhibitory interneurons that mediate lateral interactions similar to those of the horizontal cells. The ganglion cells are the output cells of the retina, and their outgoing axons form the **optic nerve**. Electrical signals in the ganglion cells are carried by action potentials, and amacrine cells also make action potentials in response to depolarization. In the other retinal neurons (photoreceptors, horizontal cells, and bipolar cells), action potentials are not involved.

Color information originates in **cone photoreceptors**. In humans, there are three types of cones: **S cones** are most sensitive to short wavelengths (which we perceive as blue), **M cones** are sensitive to the middle wavelengths (green and yellow), and **L cones** are most sensitive to long wavelengths (red). Some types of ganglion cell have receptive fields that originate from particular cone types, and these ganglion cells are sensitive to the wavelength of illumination. The **color-sensitive ganglion cells** have center-surround receptive fields, but the center and surround regions originate from different types of cone. For this reason, lights in different parts of the visual spectrum produce opposing effects on color-sensitive ganglion cells.

Topic Test 2: Synaptic Organization of the Retina

True/False

1. The second-order neurons of the retina are horizontal cells and amacrine cells.

2. The two types of lateral interneurons of the retina are horizontal cells and amacrine cells.

3. Like photoreceptors, horizontal cells are depolarized in darkness and hyperpolarize in response to illumination.

4. All bipolar cells are depolarized in darkness and hyperpolarize in response to illumination.

5. Both bipolar cells and ganglion cells have center-surround receptive fields.

6. The optic nerve consists of the axons of ganglion cells of the retina.

Multiple Choice

7. At dim light levels (e.g., at night with a full moon), we are unable to distinguish colors of objects because
   a. only a single type of cone is stimulated by dim light.
   b. dim light does not contain the full spectrum of wavelengths within the visible spectrum.
   c. only rod photoreceptors can detect dim light, but stimulation of the cones is required for color vision.
   d. dim light stimulates the photoreceptors but not the ganglion cells.
   e. only cones are stimulated by dim light, but stimulation of both cones and rods is required for color vision.

8. Suppose a ganglion cell has a center-surround receptive field, with an excitatory center. Which of the following response patterns would be observed?
   a. A light spot on a dark background, with the spot covering the center of the receptive field, would excite the cell.
   b. A dark spot on a bright background, with the spot covering the center of the receptive field, would inhibit the cell.
   c. A uniform bright light covering both the center and surround of the receptive field would have little effect on firing rate.
   d. All of the above.

Short Answer

9. How many different types of photoreceptor are found in the human retina, and how are they distinguished?

10. List the five general classes of neurons found in the vertebrate retina.

Topic Test 2: Answers

1. **False.** The second-order cells are the cells that receive synaptic inputs from the primary receptors, the photoreceptors. Horizontal cells and bipolar cells receive synaptic inputs from the photoreceptors, but amacrine cells do not. Instead, amacrine cells receive synaptic inputs from bipolar cells. Thus, amacrine cells are third-order cells.

2. **True.** Horizontal cells make lateral synaptic connections at the level of the synaptic terminals of photoreceptors. This layer of the retina, where there are extensive synaptic connections among the photoreceptor, bipolar cells, and horizontal cells, is called the outer plexiform layer. Amacrine cells are also lateral interneurons, but they make their synaptic connections at the level of the synaptic terminals of the bipolar cells. This layer of the retina is called the inner plexiform layer, where synaptic interactions occur among bipolar cells, amacrine cells, and ganglion cells.

3. **True.** Glutamate is released from the photoreceptor synapse in darkness, when photo-receptors are depolarized. Glutamate-gated nonspecific cation channels are opened in the horizontal cells in darkness, which depolarizes the horizontal cell. Illumination decreases glutamate release from the photoreceptor synapse, and the membrane potential of the horizontal cell hyperpolarizes.

4. **False.** There are two types of bipolar cell in the retina, distinguished on the basis of their response to light. Off-type bipolar cells are depolarized in darkness and hyper-polarize in response to light. However, on-type bipolar cells are hyperpolarized in darkness and depolarize in response to light. In both cases, the light response of the bipolar cell is caused by decreased release of glutamate by photoreceptors. In off-type bipolar cells, glutamate directly opens nonspecific cation channels. In on-type bipolar cells, glutamate activates a G protein-coupled receptor. The activated G protein in turn activates phosphodiesterase, reducing cyclic GMP in the bipolar cell in darkness. As in photoreceptors, cyclic GMP opens a nonspecific cation channel. In darkness, presynaptic glutamate release is high, postsynaptic cyclic GMP is low, and the cation channels are closed. In the light, glutamate release is low, cyclic GMP levels rise, and channels open, depolarizing the bipolar cell.

5. **True.** Lateral inhibitory interactions mediated by synapses of the horizontal cells form the basis for center-surround receptive fields of bipolar cells. Lateral inhibitory interactions mediated by the amacrine cells contribute to center-surround receptive fields of ganglion cells. Some ganglion cells are excited by light placed in the center of the receptive field (on-type, or on-center ganglion cells) and inhibited by light placed in the surround, whereas other ganglion cells are inhibited by light placed in the center of the receptive field (off-type, or off-center ganglion cells) and excited by light in the surround.

6. **True.** Axons of ganglion cells cross over the inner surface of the retina in a layer called the nerve fiber layer. The axons converge at a single exit point (the optic disk) and form the optic nerve.

7. **c.** Cone photoreceptors are necessary for color vision, but they are relatively insensitive to light. The cones detect light at relatively high intensities, such as daylight or normal room lighting.

8. **d.** With a center-surround receptive field, responses elicited by stimuli applied in the center and surround antagonize each other. For an on-center cell, the optimal excitatory stimulus would be a light that just fills the central portion of the receptive field without extending into the surround, and the optimal inhibitory stimulus would be a light that fills the concentric surround region without extending into the center. Light falling uniformly in both parts of the receptive field would activate competing excitatory and inhibitory synaptic influences and so would have little effect on the firing rate.

9. Human beings have four types of photoreceptor cell: three types of cone and a single type of rod. Each type of photoreceptor has a different visual pigment molecule, characterized by a different opsin protein. Each visual pigment has different spectral sensitivity to light. S cones absorb light best at a wavelength of about 420 nm, M cones at about 530 nm, and L cones at about 560 nm. The rod photopigment rhodopsin is maximally sensitive to light at about 500 nm.

10. The five neuron classes are photoreceptors, horizontal cells, bipolar cells, amacrine cells, and ganglion cells.

TOPIC 3: HIGHER VISUAL PROCESSING

KEY POINTS

✓ *What parts of the brain are involved in processing visual information?*

✓ *What part of the thalamus receives input from the retina?*

✓ *Where in the cerebral cortex is the primary visual cortex located?*

✓ *What is the functional organization of the visual cortex?*

The optic nerves from the two eyes enter the skull and merge at the **optic chiasm** (**Figure 8.2A**). At the chiasm, axons coming from ganglion cells in the nasal half of each retina cross to the opposite of the brain, whereas axons coming from ganglion cells in the temporal half of each retina remain on the same side of the brain. Thus, the left half of the visual field is represented in the right half of the brain and the right visual field in the left half of the brain (**Figure 8.2B**). The synaptic targets of the retinal ganglion cells in the brain include the **optic tectum** and, in mammals, the **lateral geniculate nucleus** of the thalamus. From the

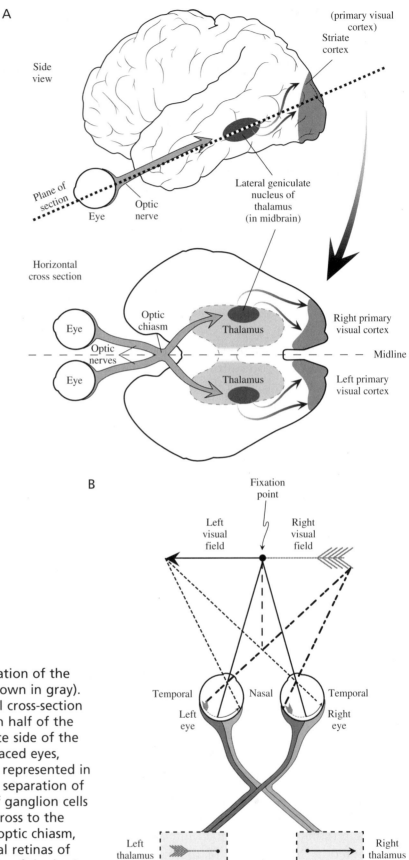

A

Side view

(primary visual cortex)
Striate cortex

Plane of section

Eye

Optic nerve

Lateral geniculate nucleus of thalamus (in midbrain)

Horizontal cross section

Eye

Optic chiasm

Optic nerves

Eye

Thalamus

Thalamus

Right primary visual cortex

Midline

Left primary visual cortex

B

Fixation point

Left visual field

Right visual field

Temporal

Left eye

Nasal

Right eye

Temporal

Left thalamus

Right thalamus

Figure 8.2 A. Anatomical organization of the geniculostriate visual pathway (shown in gray). Top: side view; bottom: horizontal cross-section in the plane shown above. B. Each half of the visual field projects to the opposite side of the brain. In animals with frontally placed eyes, the left and right visual fields are represented in both eyes. To produce the correct separation of the two visual fields, the axons of ganglion cells in the nasal part of both retinas cross to the opposite side of the brain at the optic chiasm, while the axons from the temporal retinas of both eyes remain on the same side of the brain.

thalamus, visual information is relayed to the **primary visual cortex** (also called area V1 or the striate cortex).

In the lateral geniculate nucleus, projection neurons have receptive fields like those of the retinal ganglion cells, with a center-surround organization. In the primary visual cortex, however, a variety of different receptive fields are observed. Many neurons in the area V1 respond best to a bar of light in a particular orientation. These orientation-selective cortical neurons are the building blocks for detecting edges that define borders of objects in the visual world. Some cortical neurons are motion sensitive and also require that the bar of light must move in a particular direction. Other cortical neurons are color sensitive.

Primary visual cortex is organized into a two-dimensional array of vertical columns that extend from the cortical surface to the underlying white matter. One dimension consists of **ocular dominance columns**, which receive inputs preferentially from either the contralateral or the ipsilateral eye. The other dimension consists of **orientation columns**, within which all the orientation-selective cells are best stimulated by lines of the same orientation but at slightly different retinal locations. Within the primary visual cortex, each point on the retinal surface is examined in parallel by orientation-selective neurons whose preferred orientations cover all directions. Similarly, each point on the retina is also examined simultaneously by motion-sensitive neurons whose preferred directions of movement cover all possibilities. The primary visual cortex thus carries out in parallel an analysis of the visual field for a wide variety of aspects of visual stimuli, including line orientation, direction of movement, and color.

Orientation-selective neurons in the primary visual cortex are used as building blocks to construct cells with more complex receptive fields. Some of the more complex cells require that lines must be a specific length, as well as orientation. Information from the primary visual cortex is passed on to higher order cortical areas for further processing. **Area V2** is found just anterior to the primary visual cortex (area V1), receives inputs from area V1, and has subdivisions devoted to object form, color, and motion. In turn, both area V1 and area V2 project to more specialized areas found in progressively more anterior portions of the occipital lobe and the posterior part of the temporal lobe. **Area V4** is specialized for color vision, **area V3** for analysis of object shape, and **area V5** for detection of motion.

Topic Test 3: Higher Visual Processing

True/False

1. The optic nerve from each eye projects to the thalamus on the contralateral side of the brain.

2. The optic chiasm is the nucleus of the thalamus that receives the inputs from the optic nerve.

3. Neurons of the lateral geniculate nucleus faithfully relay the output of the retinal ganglion cells to the cortex, without modification.

4. Like the retinal ganglion cells, the neurons of the primary visual cortex have center-surround receptive fields.

5. In the primary visual cortex, the inputs from the two eyes are segregated into vertically oriented columns running perpendicular to the cortical surface.

6. The primary visual cortex (area V1) is located in the most posterior portion of the occipital cortex, at the posterior pole of the cortex.

7. The higher order visual cortical regions (areas V2, V3, V4, and V5) are located in progressively more anterior portions of the cortex in the occipital lobe and posterior part of the temporal lobe.

Multiple Choice

8. Which of the following describes a receptive field that might be observed in area V1?
   a. a center-surround receptive field, with color sensitivity within both the center and the surround portions of the receptive field.
   b. an oblong bar-shaped receptive field, in which a line of light of a specific width and orientation produces excitation when placed at a specific location within the visual field.
   c. a large receptive field, covering a substantial portion of the visual field, in which a line of light of a specific orientation produces excitation when placed anywhere within the receptive field.
   d. all of the above.

9. The thalamus on the right side of the brain receives inputs from
   a. ganglion cells in the nasal half of the right retina and the temporal half of the left retina.
   b. ganglion cells in the temporal half of the right retina and the nasal half of the left retina.
   c. neurons of the primary visual cortex on the right side of the brain.
   d. b and c.

Short Answer

10. The part of the thalamus that receives visual input from the retina is the _____ (three words).

Topic Test 3: Answers

1. **False.** The left half of the visual field projects to the right thalamus, and the right half of the visual field projects to the left thalamus. However, both visual fields are represented in both eyes (see Figure 8.2B). Light coming from the left half of the visual field falls onto the nasal half of the retina in the left eye and the temporal half of the retina in the right eye. The reverse is true for light coming from the right half of the visual field. To produce the proper sorting of visual field halves, axons from ganglion cells in the temporal half of each retina remain on the same side of the brain, whereas axons from ganglion cells in the nasal half cross to the other side of the brain at the optic chiasm.

2. **False.** The optic chiasm is the crossover point at the base of the brain where the two optic nerves come together.

3. **False.** The lateral geniculate nucleus receives inputs from the reticular formation and also a substantial synaptic feedback from the primary visual cortex. These inputs alter the ability of the thalamic neurons to pass visual information on to the cortex. The thalamus also includes local interneurons that modify the responses of the relay neurons to visual stimuli.

4. **False.** Receptive fields of neurons in the primary visual cortex are complex and varied. The cells commonly respond best to bright or dark lines rather than to simple dots.

5. **True.** The thalamus on each side of the brain receives synaptic inputs from both eyes. In the lateral geniculate nucleus, axonal inputs are spatially segregated into different layers of the nucleus. In the primary visual cortex, the inputs remain segregated, but neurons receiving input from each eye are organized into vertical columns, called ocular dominance columns. The columns from ipsilateral and contralateral eyes alternate in a regular array across the cortex.

6. **True.** The primary visual cortex receives the thalamic inputs from the lateral geniculate nucleus. It is located in the most posterior part of the occipital lobe at the back of the cerebrum. Other names for the primary visual cortex are area V1, Brodmann's area 17, and the striate cortex.

7. **True.** The second level of visual processing in the cortex occurs in area V2, which receives extensive inputs from cortical neurons in area V1. Area V2 is also located in the occipital lobe, just in front of the primary visual cortex. Moving progressively forward through the cortex, we next encounter area V3, then V4, and finally V5. The higher order areas are mostly buried within the deep infoldings of the cortical surface.

8. **d.** Receptive fields in primary visual cortex are quite varied. Color-sensitive cells described in statement a are sometimes observed. The situation described in statement b occurs for orientation-selective cortical neurons called simple cells. These cells respond best to lines of a particular orientation located at a particular position on the retina. Statement c describes the response of cortical neurons called complex cells, which respond best to stimuli of a particular orientation, independent of stimulus location within broad limits.

9. **d.** Neurons in the lateral geniculate nucleus receive several inputs related to visual processing. Inputs from the retinal ganglion cells are as described in statement b: Temporal ganglion cells project to the same side of the brain and nasal ganglion cells project to the contralateral thalamus. In addition, neurons in the primary visual cortex send axons back to the thalamus, where they make excitatory synapses onto thalamic projection neurons and onto local inhibitory interneurons within the thalamus.

10. Lateral geniculate nucleus.

IN THE CLINIC

A variety of diseases affect the visual system, and neurons of the visual system are targets for many of these diseases. Because the retina can be directly observed through the pupil of the eye, many diseases affecting the retina can be diagnosed in routine eye examinations. Of course, functional photoreceptors are required for vision, and a number of hereditary ailments produce blindness by causing degeneration of the photoreceptors. In many of these hereditary diseases, mutated genes underlying the disease have been identified, and molecules of phototransduction are frequently the products encoded by the genes. For example, photoreceptor degeneration is known to result from mutations in genes for visual pigment molecules, phosphodiesterase, and cyclic GMP-gated ion channels. Other diseases (e.g., glaucoma) affect vision by loss of retinal ganglion cells, whose axons form the output path from the retina. Strokes in the occipital lobe also produce visual deficits, ranging from complete loss of conscious visual perception if the primary visual cortex is affected to more discrete perceptual disturbances if selected higher order visual areas are affected.

Chapter Test

True/False

1. The ion channels underlying the light response of photoreceptors are directly gated by cyclic GMP.

2. The optic nerve carries the axons of photoreceptors from the eye to the retinal ganglion cells of the thalamus.

3. Light is absorbed in photoreceptors by visual pigment molecules, which consist of opsin protein molecules linked to a light-absorbing chromophore group (retinal).

4. Absorption of light causes photoreceptors to hyperpolarize, which reduces the release of glutamate from the photoreceptor synaptic terminal.

5. Cone photoreceptors make synapses onto horizontal cells but not bipolar cells, whereas rod photoreceptors make synapses onto bipolar cells but not horizontal cells.

6. In the retina, only photoreceptors have receptive fields because they are the only cells that directly respond to light.

7. In the visual system, center-surround receptive fields are first observed in retinal ganglion cells.

8. The lateral geniculate nucleus on each side of the brain receives inputs from both eyes.

9. The primary visual cortex is located in the posterior part of the temporal lobe.

10. Some neurons in the primary visual cortex respond best to lines of a particular orientation.

Multiple Choice

11. Lateral inhibitory neurons of the retina are
    a. amacrine cells and bipolar cells.
    b. horizontal cells and amacrine cells.
    c. bipolar cells and ganglion cells.
    d. photoreceptor cells and ganglion cells.

12. Which of the following is not associated with the visual system?
    a. thalamus
    b. striate cortex
    c. rhodopsin
    d. transducin
    e. postcentral gyrus

13. Which of the following is not part of the phototransduction process?
    a. 11-cis retinal
    b. opsin
    c. phosphodiesterase
    d. glutamate
    e. cyclic GMP

14. The lateral geniculate nucleus receives synaptic inputs from which of the following sources?

a. retinal ganglion cells
b. primary visual cortex
c. optic nerve
d. retina
e. all of the above

15. The four types of photoreceptor found in the human retina are
a. rods, S cones, M cones, and L cones.
b. rods, cones, bipolar cells, and horizontal cells.
c. rods, ganglion cells, lateral geniculate neurons, and cortical neurons.
d. rhodopsin, opsin, all-trans retinal, and transducin.

Short Answer

16. What are the two types of retinal neuron that receive synaptic inputs from photoreceptors? What light responses are observed in these second-order neurons?

17. Matching:
area V3   area V4   area V5
color vision   analysis of object shape   motion perception

18. Identify the following: orientation column, lateral geniculate nucleus, optic chiasm.

Essay

19. Briefly describe the sequence of events leading from absorption of light to a change in permeability of the photoreceptor.

20. The processing of visual information in the brain is both parallel and hierarchical. Explain whether this statement is accurate or inaccurate.

Chapter Test Answers

1. **T** 2. **F** 3. **T** 4. **T** 5. **F** 6. **F** 7. **F** 8. **T** 9. **F** 10. **T** 11. **a** 12. **e**

13. **d** 14. **e** 15. **a**

16. Horizontal cells and bipolar cells. Horizontal cells hyperpolarize in response to light. Bipolar cells are of two types, both with center-surround receptive fields: On-type bipolar cells depolarize in response to light in the center of the receptive field and off-type bipolar cells hyperpolarize in response to light in the center of the receptive field.

17. area V3, analysis of object shape; area V4, color vision; area V5, motion perception.

18. Orientation column: a vertically oriented column within the primary visual cortex. Within an orientation column, all of the cortical neurons have the same orientation specificity for bar-shaped stimuli. Lateral geniculate nucleus: part of the thalamus that receives inputs from retinal ganglion cells via the optic nerve. Optic chiasm: structure at the base of the brain where the two optic nerves come together. At the chiasm, axons from ganglion cells of the nasal retina cross over to the opposite side of the brain.

19. Light is absorbed by 11-cis retinal, which undergoes photoisomerization to all-trans retinal. This induces a conformation change in opsin, which is then able to interact

with and activate the G protein transducin. Activated transducin in turn activates phosphodiesterase, which hydrolyzes cyclic GMP to GMP, reducing cyclic GMP in the photoreceptor. Cyclic GMP directly binds to and opens nonspecific cation channels in the plasma membrane, and so the fall in cyclic GMP concentration closes the channels.

20. Information processing in the visual system is both parallel and hierarchical. Many different types of neuron in the visual cortex examine the same point in the visual field simultaneously, each type analyzing a particular aspect of the visual stimulus at the same time. This is parallel processing. For example, within an orientation column in area V1, the orientation-sensitive neurons simultaneously examine a small portion of the visual field for lines of their preferred orientation. In the neighboring orientation column, the cells examine that same portion of the visual field for lines of a slightly different orientation and so on through the array of orientation columns. In the cortex as a whole, all possible orientations are represented at all positions within the visual field. The same holds true for motion-sensitive cells, color-sensitive cells, and so on. Information processing in the visual system is also hierarchical. At each level, neurons concerned with a particular aspect of vision feed cells at the next higher level, which then carry the analysis further. For example, cells of area V2 concerned with line orientation receive inputs from orientation-selective neurons of area V1, and neurons of area V3, which are involved in form vision, in turn receive inputs from the orientation-selective cells of area V2.

Check Your Performance:

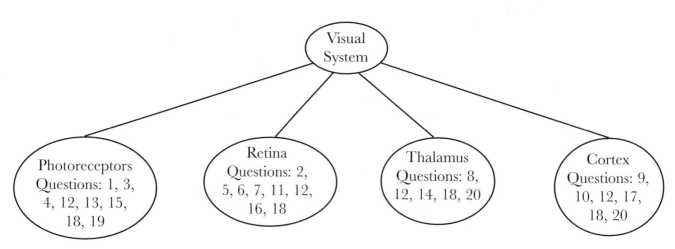

Note the number of questions in each grouping that you got wrong on the chapter test. Identify areas where you need further review and go back to relevant parts of this chapter.

Further help: If you continue to have difficulty with the concepts, a detailed presentation of these topics can be found in chapters 16 and 17 of *Neurobiology: Molecules, Cells, and Systems*, published by Blackwell Science, Inc.

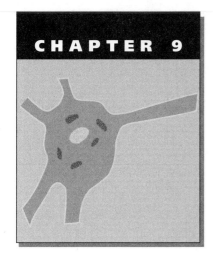

CHAPTER 9

The Auditory System

In mammals, the sense of hearing (**audition**) is the principal vibration sense, and the sensory organ is the ear. This chapter examines the sensory receptor cells of the auditory system and the brain mechanisms that analyze the sensory information provided by the ear.

ESSENTIAL BACKGROUND

- **The physical nature of sound stimuli (sinusoidal oscillations of pressure)**
- **The anatomy of the ear**
- **The anatomical organization of the brain stem, midbrain, and cerebral cortex**
- **Ionic permeability and membrane potential**

TOPIC 1: HAIR CELLS AND THE COCHLEA

KEY POINTS

✓ *What are the mechanoreceptors of the sense of hearing?*

✓ *How do the mechanoreceptors transduce vibration into an electrical signal?*

✓ *How is the frequency of vibration encoded in the cochlea of the ear?*

Vibrations of the air or water in which animals live are sensed by a special type of mechanoreceptor, **hair cells**, which have a bundle of modified cilia at one end. The hair cell translates movements of the cilia into electrical signals. The cilia are tethered to one another via filamentous threads near their tips, and when the bundle is deflected toward the longest cilium, the filaments exert tension on cation channels in the cilia, opening the ion channels and depolarizing the cell. Deflection in the opposite direction closes the channels and hyperpolarizes the cell. If the hair bundle oscillates back and forth, the membrane potential of the hair cell also oscillates in phase with the movements (**Figure 9.1**). The hair cell makes an excitatory synapse onto a sensory neuron, which reports the activity of the hair cell to the nervous system.

In the mammalian ear, hair cells are located in the **cochlea**, a spiral-shaped compartment in the inner ear. Vibrations in the air are transferred to the fluid inside the cochlea via the **tympanic membrane** and the bones of the middle ear. Vibrations move the **basilar membrane**, which has the hair cell-containing **organ of Corti** riding on top. Motions of the basilar membrane stimulate the cilia of the hair cells, which make excitatory synapses on the sensory neurons of

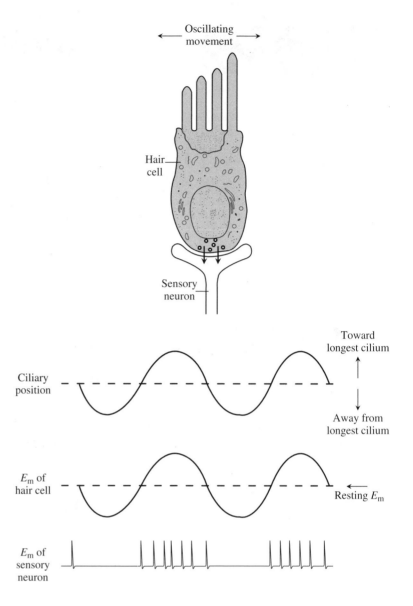

Figure 9.1 Oscillatory motion of the ciliary bundle produces oscillating changes in membrane potential in the hair cell. These oscillatory changes are reflected in bursts of action potential in the sensory neuron, where the bursts occur with the same periodicity as the oscillating stimulus.

the **spiral ganglion**. The axons of the spiral ganglion neurons make up the auditory nerve and carry auditory information into the brain.

Each auditory nerve fiber is sensitive to sound stimuli within a specific frequency range, which depends on the part of the cochlea the nerve fiber innervates. Low-frequency sounds (<100 Hz) are sensed at the apical end of the cochlea, and high-frequency sounds (up to 20,000 Hz in humans) are sensed at the basal end. Frequency tuning of the cochlea is determined by mechanical properties of the basilar membrane, by electrical tuning of individual hair cells, and by active motions of hair cells.

Topic Test 1: Hair Cells and the Cochlea

True/False

1. Hair cells are sensory neurons that innervate the hair follicles of the skin.

2. Hair cells depolarize when the ciliary bundle is deflected toward the longest cilium in the bundle.

3. Each sensory neuron in the spiral ganglion responds equally well to all sound frequencies.

4. Individual hair cells have a preferred stimulus frequency and show the largest oscillation in membrane potential when the cilia are oscillated at this preferred frequency.

5. The hair cells of the cochlea are in the organ of Corti.

6. Hair cells move in response to changes in membrane potential.

Multiple Choice

7. Hair cells of the cochlea
   a. have axons that project into the brain via the auditory nerve.
   b. fire action potentials in response to depolarization and cease firing action potentials in response to hyperpolarization.
   c. receive synapses from incoming nerve fibers and make synapses onto sensory neurons of the spiral ganglion.
   d. none of the above.

8. Frequency tuning in the cochlea arises from
   a. the shape and stiffness of the basilar membrane.
   b. the electrical properties of individual hair cells.
   c. movements produced by the hair cells.
   d. all of the above.

Short Answer

9. Identify spiral ganglion.

10. Identify tectorial membrane.

Topic Test 1: Answers

1. **False.** Hair cells are ciliated mechanoreceptors that respond to deflection of the bundle of cilia. Hair cells are found in a variety of sensory organs, including the lateral line system of fish and amphibians, vestibular organs, otolith organs, and the cochlea.

2. **True.** Cilia of hair cells increase progressively in length from one end of the bundle to the other. The cilia are tied together near their tips by fine filaments that extend from each cilium to the next. The filaments transmit tension to their attachment points. Increased tension opens cation ion channels in the cilia. When the cilia move toward the longest cilium, tension increases, and when the cilia move in the opposite direction, tension decreases.

3. **False.** Human listeners can detect sound in the frequency range from 20 Hz to approximately 20,000 Hz. Each nerve fiber in the auditory nerve, however, is most sensitive to a specific frequency within this range. Each neuron in the spiral ganglion receives synaptic input from hair cells in a particular part of the cochlea, and each part of the cochlea responds best to a particular frequency.

4. **True.** Some hair cells respond best to low-frequency stimuli, whereas others respond best to high-frequency stimuli. Tuning of a hair cell to a specific frequency results from the interplay of two types of ion channel in the hair cell: voltage-activated Ca^{2+} channels and calcium-activated K^+ channels. Depolarization opens Ca^{2+} channels and increases intracellular Ca^{2+} concentration. Increased Ca^{2+} opens calcium-activated K^+ channels, which hyperpolarize the cell and close Ca^{2+} channels. As internal Ca^{2+} declines again, K^+ channels close and the cell depolarizes once more, opening Ca^{2+} channels and repeating the cycle. In this manner, the membrane potential of the hair cell oscillates at a frequency that depends on the delay time between the opening of each type of channel. The delay time is short in high-frequency cells and long in low-frequency cells.

5. **True.** The basilar membrane forms the barrier between two of the three compartments of the cochlea (the scala tympani and the scala media). The organ of Corti rests on top of the basilar membrane, within the scala media. The hair cells are arranged in orderly rows in the organ of Corti.

6. **True.** Hair cells produce two motions in response to changes in membrane potential. First, the ciliary bundle moves back and forth and the hair cell expands and contracts in length in response to oscillations of membrane potential. Active movements of hair cells play an important role in amplifying small motions of the basilar membrane.

7. **c.** In addition to synapses made by hair cells onto outgoing sensory fibers (afferent synapses), hair cells also receive synaptic inputs (efferent synapses) from a different set of nerve fibers. The incoming axons originate from neurons in the brain stem. The efferent synaptic terminals release the neurotransmitter acetylcholine, which activates G protein-coupled receptors in the hair cells and indirectly opens K^+ channels. Thus, activation of the efferent input hyperpolarizes the hair cells and inhibits the auditory transduction process.

8. **d.** Statement a is correct because the basilar membrane is tapered in width along its length, with the apical end being wider and the basal end narrower. Like the strings of a harp, the short end of the membrane vibrates best at high frequencies, and the long end vibrates best at low frequencies. In addition, the basal end of the basilar membrane is stiffer than the apical end, which adds to the preferential vibration of the basal end at high frequencies. Statement b is correct because each hair cell has a preferred stimulus frequency. Hair cells at a particular position along the basilar membrane have a preferred frequency that matches the frequency tuning of the membrane at that position. Thus, the electrical properties of the hair cells and the mechanical properties of the basilar membrane reinforce each other in establishing frequency tuning. Statement c requires further explanation about the structure of the organ of Corti. The organ of Corti contains two groups of hair cells, the inner hair cells and the outer hair cells. The outer hair cells are more numerous than the inner hair cells (approximately 20,000 outer hair cells and 4,000 inner hair cells in the entire cochlea), are larger than the inner hair cells and have larger hair bundles. The primary function of the outer hair cells is to amplify motion of the basilar membrane by means of the active movements described above.

9. The spiral ganglion is located within the cochlea. Neurons of the spiral ganglion send axon branch into the organ of Corti and another branch into the brain via cranial nerve VIII (the auditory nerve). Spiral ganglion neurons receive excitatory synapses from auditory hair cells.

10. The tectorial membrane extends over the top of the organ of Corti, inside the middle chamber of the cochlea (the scala media). Ciliary bundles of outer hair cells are embedded in the tectorial membrane.

TOPIC 2: THE AUDITORY SYSTEM OF THE BRAIN

KEY POINTS

✓ *What part of the brain receives input from the auditory nerve?*

✓ *What part of the thalamus receives auditory information?*

✓ *Where is the primary auditory cortex located?*

Auditory nerve fibers terminate in the **cochlear nucleus** in the brain stem. This nucleus relays specific information about the auditory stimulus to the **superior olivary nucleus** and the **inferior colliculus**. The superior olivary nucleus is concerned with localization of sounds with respect to the head, which is based on differences in arrival time and sound intensity in the two ears. From the inferior colliculus, auditory information is sent to the **medial geniculate nucleus** of the thalamus, which in turn relays auditory signals to the primary auditory cortex in the superior portion of the temporal lobe.

Primary auditory cortex is organized into a two-dimensional array of vertical cell columns. One dimension consists of frequency columns, within which neurons have the same preferred stimulus frequency. The other dimension consists of binaural columns, in which the cells respond best either to binaural or to monaural stimulation.

Topic Test 2: The Auditory System of the Brain

True/False

1. Auditory nerve fibers project directly to the thalamus, without intervening synapses.

2. The first auditory relay station in the brain is the cochlear nucleus.

3. The cochlear nucleus contains a spatial representation of sound frequency, called a tonotopic map.

4. All neurons in the cochlear nucleus respond in the same way to auditory stimuli.

5. Neurons of the cochlear nucleus project exclusively to the inferior colliculus on the same side of the brain.

6. The primary auditory cortex is located in the superior portion of the occipital cortex.

Multiple Choice

7. The part of the thalamus involved in processing auditory information is
   a. the lateral geniculate nucleus.
   b the lateral lemniscus.
   c. the medial geniculate nucleus.
   d. the reticular formation.

8. The primary auditory cortex is organized into a two-dimensional array of vertically oriented columns:
   a. binaural columns and monaural columns.
   b. frequency columns and temporal columns.
   c. summation columns and suppression columns.
   d. frequency columns and binaural columns.

Short Answer

9. Describe the synaptic connections (inputs and outputs) of the superior olivary nucleus.

10. What synaptic arrangement accounts for the fact that auditory neurons in the thalamus and cortex respond to sound stimuli in both ears.

Topic Test 2: Answers

1. **False.** Incoming auditory nerve fibers terminate in the brain stem, in the cochlear nucleus. A minimum of two synapses is required for information from the spiral ganglion to reach the thalamus.

2. **True.** The cochlear nucleus is located in the brain stem. The axons of the spiral ganglion cells in the cochlea on each side of the head terminate in the cochlear nucleus on the same side of the brain.

3. **True.** The auditory nerve fibers originating from different parts of the cochlea are geometrically mapped onto specific regions within the cochlear nucleus. Because each position along the cochlea is most sensitive to a particular stimulus frequency, this spatial map corresponds to a map of preferred stimulus frequency. Thus, a stimulus of a particular pitch will stimulate cochlear nucleus neurons at a specific location. The spatial map of preferred frequency is called a tonotopic map.

4. **False.** Some neurons in the cochlear nucleus fire only a single action potential at the onset of a tone. These neurons extract precise information about the time of onset of an arriving sound. Other cochlear nucleus neurons remain silent at the onset of a stimulus and then increase their firing rate during a sustained tone. These cells transmit information about sound intensity.

5. **False.** Although the ipsilateral inferior colliculus is one target of the cochlear nucleus, it is not the exclusive target. Information from the cochlear nucleus is also sent to the superior olivary nucleus on both sides of the brain and to the inferior colliculus on the contralateral side. Because of the bilateral projections from the cochlear nucleus, neurons in the inferior colliculus and superior olivary nucleus receive synaptic inputs from both ears.

6. **False.** The primary auditory cortex is actually located in the superior part of the temporal lobe.

7. **c.** The medial geniculate nucleus is the part of the thalamus that receives auditory inputs. The medial lemniscus (statement b) is in fact part of the auditory system, but it is the fiber tract carrying axons from the cochlear nucleus and superior olivary nucleus to the inferior colliculus.

8. **d.** Primary auditory cortex has a two-dimensional array of functionally defined columns running vertically from the surface of the cortex to the underlying white matter. One dimension of the array consists of frequency columns, within which cells have the same frequency preference for sound stimuli. The other axis of the array consists of binaural columns. There are two types of binaural columns, summation columns and suppression columns, that alternate along the surface of the cortex. In summation columns, the greatest response is observed when a sound is presented simultaneously to both ears. In suppression columns, neurons respond best when a sound is presented to only one ear. Frequency columns are analogous to orientation columns of visual cortex, and binaural columns are analogous to ocular dominance columns of visual cortex.

9. The superior olivary nucleus receives inputs from the cochlear nuclei on both sides of the brain. The neurons of the superior olivary nucleus have two synaptic targets. First, axons ascend to the inferior colliculus on both sides of the brain. These ascending axons travel in the medial lemniscus, together with ascending axons from the cochlear nucleus. Second, neurons in the superior olivary nucleus are the source of the efferent axons that project back to the cochlea via the auditory nerve, where they make synapses on the outer hair cells.

10. Ascending axons from the cochlear nuclei project to the inferior colliculi on both sides of the brain. Therefore, neurons in the inferior colliculus receive binaural inputs. Also, ascending connections from the superior olivary nucleus (which itself receives inputs from both cochlear nuclei) project to the inferior colliculi on both sides of the brain. Connections from the inferior colliculus to the thalamus and from the thalamus to the cortex are largely ipsilateral in the auditory system, but the binaural inputs are already established at the level of the outputs from the cochlear nuclei and the superior olivary nuclei.

IN THE CLINIC

Because the ear is a mechanical system, a common source of hearing loss is mechanical damage to the auditory apparatus, often caused by exposure to loud sounds. When motions of the basilar membrane are too large, hair cells can be damaged or destroyed or ciliary bundles of outer hair cells can be dislodged from the tectorial membrane. If this happens, sensitivity is reduced at the frequency corresponding to the damaged portion of the cochlea. In addition to the loss of sensitivity, the remaining hair cells in the damaged area may become spontaneously active. This may occur because of the electrical tuning mechanisms of the hair cell, which can produce spontaneous oscillations of membrane potential at the preferred frequency of the hair cell even in the absence of stimulation. Spiral ganglion neurons receiving inputs from spontaneously active hair cells fire action potentials just as they would in response to sound, and so the auditory system interprets the spontaneous activity as a continuous sound. The result is "ringing in the ears" or tinnitus, which can be more debilitating for patients than the actual hearing loss itself.

Chapter Test

True/False

1. The mechanosensory cells of the ear are hair cells.

2. The basilar membrane is mechanically tuned so that sounds of different frequencies cause maximal vibration of the membrane at different locations.

3. Each spiral ganglion neuron fires action potentials at the highest rate at a particular preferred sound frequency.

4. At every longitudinal position along the cochlea, all hair cells respond best to high-frequency sound (>10,000 Hz).

5. The electrical tuning of individual hair cells to a particular preferred frequency depends on the interplay between neurotransmitter released from synapses made by the hair cell

onto afferent fibers and neurotransmitter released from synapses made onto the hair cell by efferent fibers.

6. The cochlear nucleus is located within the cochlea and contains neurons that receive synapses from hair cells.

7. Neurons of the superior olivary nucleus are concerned with localization of a sound source relative to the head.

8. Neurons of the medial geniculate nucleus of the thalamus on each side of the brain receive auditory information only from the contralateral ear.

9. A tonotopic map is present in the primary auditory cortex but not in the other brain nuclei concerned with auditory processing.

10. Neurons of the cochlear nucleus make synapses on neurons of the inferior colliculus and superior olivary nucleus.

Multiple Choice

11. In the hair cells of the cochlea,
    a. deflection of the ciliary bundle toward the longest cilium depolarizes the cell.
    b. deflection of the ciliary bundle away from the longest cilium hyperpolarizes the cell.
    c. the change in membrane potential upon hair bundle deflection is caused by opening and closing mechanosensitive ion channels near the tips of the cilia.
    d. all of the above.
    e. none of the above.

12. Which of the following is not a difference between the inner and outer hair cells of the cochlea?
    a. Outer hair cells are more numerous than inner hair cells.
    b. Most spiral ganglion neurons receive synaptic inputs from outer hair cells.
    c. The ciliary bundles of outer hair cells are embedded in the tectorial membrane.
    d. The ciliary bundles of inner hair cells do not directly contact the tectorial membrane.
    e. Active motions of outer hair cells amplify motions of the basilar membrane.

13. In the cochlea,
    a. inner hair cells are located in the organ of Corti but outer hair cells are not.
    b. the organ of Corti rests on top of the tectorial membrane.
    c. there is an orderly spatial array of preferred sound frequency, with high frequency sounds preferred at the basal end and low frequency sounds preferred at the apical end.
    d. active movements produced by outer hair cells dampen motions of the basilar membrane and reduce sensitivity to sound.
    e. only outer hair cells make synapses onto spiral ganglion neurons.

14. The primary auditory cortex
    a. is located in the superior portion of the temporal cortex on each side of the brain.
    b. receives direct input from the medial geniculate nucleus of the thalamus on the same side of the brain.

c. contains a tonotopic map of sound frequency.

d. contains some neurons that fire most strongly when a stimulus is presented to only one ear but not both.

e. all of the above

15. The inferior colliculus

a. receives direct synaptic input from both the cochlear nucleus and the superior olivary nucleus.

b. is part of the thalamus, in the diencephalon.

c. contains neurons that project directly to the primary auditory cortex, without an intervening synapse.

d. contains neurons that respond to sound only in the ipsilateral ear.

e. none of the above

16. Deafness could result from

a. destruction of the neurons of the spiral ganglion.

b. destruction of cranial nerve VIII.

c. a lesion that destroys the medial geniculate nucleus.

d. a and c, but not b.

e. a, b, and c.

Short Answer

17. List three mechanisms that contribute to the frequency tuning of single nerve fibers in the auditory nerve.

18. Identify the following: tonotopic map, summation column, cochlear nucleus.

Essay

19. Describe the most direct pathway leading from the ear to the cerebral cortex.

20. Describe the two-dimensional array of vertical columns found in the primary auditory cortex.

Chapter Test Answers

1. **T** 2. **T** 3. **T** 4. **F** 5. **F** 6. **F** 7. **T** 8. **F** 9. **T** 10. **T** 11. **d** 12. **b**

13. **c** 14. **e** 15. **a** 16. **e**

17. The three mechanisms of frequency tuning are mechanical tuning of the basilar membrane, electrical tuning produced by the interplay of voltage-activated Ca^{2+} channels and calcium-activated K^+ channels in the hair cells, and active movements of outer hair cells, which amplify vibrations of the basilar membrane.

18. Tonotopic map: orderly spatial separation of neurons with different preferred frequencies of sound stimuli within an auditory nucleus. For example, axons of spiral ganglion neurons from different parts of the cochlea project in an orderly way onto the neurons of the cochlear nucleus, producing a tonotopic map within the cochlear nucleus. Summation column: one of the two types of binaural column that form one axis of the two-dimensional array of cortical columns in the primary auditory cortex. In a summation

column, neurons fire best to sounds in both ears. Cochlear nucleus: nucleus in the brain stem that receives synaptic inputs from spiral ganglion axons entering in the auditory nerve.

19. The most direct path from the ear to the auditory cortex involves five synapses. The hair cells of the cochlea make synapses onto neurons of the spiral ganglion within the cochlea. The axons of the spiral ganglion neurons form the auditory nerve (cranial nerve VIII), which synapses onto neurons in the cochlear nucleus. Cochlear nucleus neurons project to the inferior colliculus, whose neurons in turn make synapses onto thalamic relay neurons in the medial geniculate nucleus. Finally, the thalamic neurons project to the primary auditory cortex in the temporal lobe.

20. Primary auditory cortex is organized into functionally defined columns, oriented vertically with respect to the cortical surface. Within a column, neurons have similar responses to auditory stimuli. One axis of the array consists of frequency columns, in which neurons have the same preferred sound frequency (e.g., 1,000 Hz). Along the frequency axis, the preferred stimulus frequency of neighboring frequency columns shifts in an orderly way. The other axis of the columnar array consists of binaural columns, which alternate successively between two different types. In summation columns, neurons are stimulated best by sounds presented to both ears (binaural stimuli). In suppression columns, stimulation in both ears is ineffective, and instead neurons respond best to stimuli presented in one ear (monaural stimuli).

Check Your Performance:

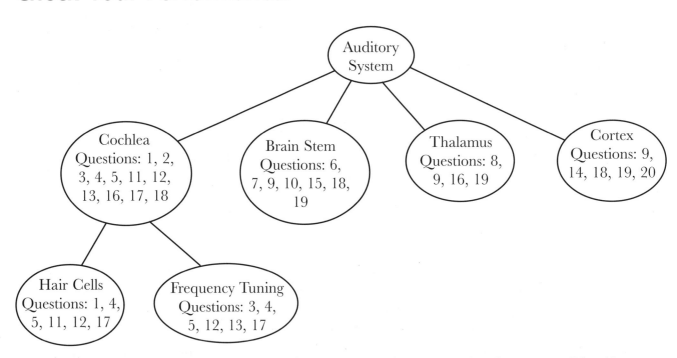

Note the number of questions in each grouping that you got wrong on the chapter test. Identify areas where you need further review and go back to relevant parts of this chapter.

Further help: If you continue to have difficulty with the concepts, a detailed presentation of these topics can be found in chapter 18 of *Neurobiology: Molecules, Cells, and Systems*, published by Blackwell Science, Inc.

CHAPTER 10

Chemical Senses

Two important chemical senses are **olfaction** (smell) and **gustation** (taste), both of which involve detection of molecules dissolved in fluid at the chemoreceptor cell's surface. This chapter examines similarities and differences in the chemotransduction mechanisms and in the processing of information in the brain for the two chemical senses.

ESSENTIAL BACKGROUND

- **Anatomy of the nose and tongue**
- **Anatomical organization of the forebrain**
- **Ionic permeability and membrane potential**
- **G proteins and second messengers**

TOPIC 1: THE OLFACTORY SYSTEM

KEY POINTS

✓ *Where are the chemoreceptor cells for olfaction?*

✓ *How do olfactory receptor cells transduce chemical signals?*

✓ *What is the olfactory bulb and how is it organized?*

✓ *What other brain regions are involved in olfactory information processing?*

For terrestrial animals, odors are transmitted via the air to the **olfactory epithelium**, located in the upper nasal passage just under the cranial cavity. Olfactory receptor cells are special ciliated neurons, whose cilia form a dense net in the layer of mucus inside the nasal passage. The cilia contain the transduction apparatus for detection of odor molecules. Olfactory receptor cells have axons that enter the brain in the **olfactory nerve** (cranial nerve I).

Odor molecules are detected by olfactory receptor molecules in the membrane of the cilia. Approximately 1,000 different types of receptor molecules exist in mammalian olfactory organs, and each type is sensitive to a particular class of odorant. Each olfactory receptor neuron is thought to express a single type of the 1,000 possible receptor molecules. Olfactory receptor molecules activate G proteins, which stimulate the synthetic enzyme adenylyl cyclase and increase cyclic AMP in the cilia. Cyclic AMP depolarizes olfactory receptor neurons by directly opening nonspecific cation channels, in a manner analogous to the action of cyclic GMP in photoreceptors. The depolarizing receptor potential initiates action potentials, which propagate along the axons of the olfactory nerve.

Axons of olfactory receptor cells terminate in the **olfactory bulbs**, which are finger-like projections of the telencephalon above the olfactory epithelia. Synaptic terminals of incoming axons cluster in spherical patches called **glomeruli** (singular: glomerulus). All olfactory receptor cells that express a particular one of the 1,000 different receptor molecule types send their axons to the same glomerulus. Thus, a single glomerulus is activated by the odor detected by a single type of olfactory receptor molecule. Within a glomerulus, axons of receptor neurons make excitatory synapses onto dendrites of **mitral cells** and **tufted cells**, which are the output neurons of the olfactory bulb. In addition, two types of lateral inhibitory neurons, **periglomerular cells** and **granule cells**, alter the activity of the output neurons.

Outputs of the olfactory bulbs project to a set of brain structures collectively called the **olfactory cortex**, including the anterior olfactory nucleus, piriform cortex, olfactory tubercle, amygdala, and entorhinal cortex. These structures are also called the paleocortex to distinguish them from the phylogenetically newer and anatomically more complex neocortex. From the paleocortex, olfactory information is sent to the medial dorsal nucleus of the thalamus, which relays olfactory signals to the frontal lobe. In addition, the paleocortex sends olfactory outputs to the limbic system, which is a complex set of interconnected brain nuclei concerned with feeding behavior, homeostasis, and emotion.

Topic Test 1: The Olfactory System

True/False

1. Olfactory receptor cells make synapses within the olfactory epithelium onto sensory neurons, which send axons into the brain.

2. Each olfactory receptor cell gives rise to numerous cilia, where the receptor molecules for olfactory stimuli are located.

3. Each olfactory receptor cell expresses only a single type of receptor molecule and thus detects a particular type of olfactory stimulus.

4. Odorant molecules directly interact with G proteins in the olfactory cilia, and the G proteins increase synthesis of cyclic AMP inside the cilia.

5. Cyclic AMP directly opens nonspecific cation channels in the olfactory cilia.

6. The olfactory bulb refers to the sensory epithelium within the nose where olfactory transduction takes place.

7. The olfactory bulb projects to the olfactory cortex, which is part of the neocortex located in the frontal lobe.

Multiple Choice

8. In the olfactory bulb,
   a. axons of olfactory receptor neurons make synapses onto dendrites of mitral cells and tufted cells.

b. tufted cells are inhibitory neurons that make inhibitory synapses onto neighboring mitral cells.

c. each glomerulus receives synaptic inputs from all of the approximately 1,000 different types of olfactory receptor cells.

d. all of the above.

e. none of the above.

9. Which of the following brain regions is not involved in processing of olfactory information?

a. piriform cortex

b. medial dorsal nucleus of the thalamus

c. limbic system

d. part of the frontal lobe of the cerebral cortex

e. all of the above are involved in the processing of olfactory information

10. In olfactory transduction,

a. the olfactory stimulus leads to activation of phosphodiesterase, which decreases cyclic AMP levels.

b. the olfactory stimulus causes the olfactory receptor cell to depolarize, which triggers action potentials in the receptor cell.

c. olfactory receptor molecules directly activate the synthetic enzyme for cyclic AMP, adenylyl cyclase.

d. all of the above.

e. none of the above.

Topic Test 1: Answers

1. **False.** The olfactory nerve contains axons of olfactory receptor cells, which directly project into the brain without intervening synapses.

2. **True.** Each olfactory receptor cell gives rise to numerous long thin cilia. The cilia are exposed to the nasal passage, and odor molecules entering the nose interact with specific receptor molecules located in the cilia.

3. **True.** In total, about 1,000 different kinds of olfactory receptor molecule exist, each one tuned to detect a particular type of odorant. Single olfactory receptor cells have only a single type of receptor molecule. So, each olfactory cell is specialized to detect a particular odorant.

4. **False.** Odor molecules actually bind to and activate receptor molecules in the olfactory cilia. The receptor molecules then activate G proteins, which stimulate synthesis of cyclic AMP by turning on the synthetic enzyme, adenylyl cyclase.

5. **True.** Olfactory chemoreception is similar to phototransduction (Chapter 8). In phototransduction, cyclic GMP directly opens nonspecific cation channels and depolarizes the photoreceptor. Olfactory transduction is similar, but the cyclic nucleotide that opens channels is cyclic AMP instead of cyclic GMP.

6. **False.** The olfactory bulb is part of the telencephalon. Axons of the olfactory nerve terminate in the olfactory bulb, which is the first stage of olfactory information processing in the brain.

7. **False.** Output neurons of the olfactory bulb do in fact project to the olfactory cortex. However, the olfactory cortex is a collective term for several brain regions, including the anterior olfactory nucleus, piriform cortex, olfactory tubercle, amygdala, and entorhinal cortex. These regions have a simpler layered structure than that found in most cortex and are collectively called the paleocortex, whereas more complex cortical areas are called neocortex.

8. **a.** The mitral and tufted cells do indeed receive synaptic inputs from the incoming axons of the olfactory receptor cells. However, both mitral and tufted cells are output neurons whose axons exit the olfactory bulb. The lateral inhibitory interneurons of the olfactory bulb are actually the periglomerular cells and the granule cells. Therefore, statement b is incorrect. Statement c is also incorrect. Each glomerulus in the olfactory bulb probably receives synaptic inputs from olfactory receptor cells of one particular type, among about 1,000 different types of receptor cell in the olfactory epithelium. Thus, cells of a single type project specifically to a particular glomerulus within the olfactory bulb.

9. **e.** The piriform cortex is part of the olfactory cortex, which receives synaptic inputs from the olfactory bulb. The medial dorsal nucleus is part of the thalamus that receives inputs from the olfactory cortex and relays the information to the neocortex in the frontal lobes. The limbic system is a diffuse set of nuclei concerned with survival functions such as feeding, drinking, sexual behavior, and emotions. The limbic system receives extensive inputs from the olfactory cortex.

10. **b.** Odor molecules combine with olfactory receptor molecules, which then stimulate G proteins. The G proteins activate adenylyl cyclase, which synthesizes cyclic AMP. Cyclic AMP opens cation channels in the olfactory cilia, leading to depolarization of the receptor cell. Depolarization stimulates action potentials, which propagate along the axon of the receptor cell into the brain. Statement a is incorrect because the enzyme activated during transduction is the synthetic enzyme for cyclic AMP, adenylyl cyclase, and the level of cyclic AMP increases rather than decreases in response to olfactory stimuli. Statement c is false because the activated receptor molecule stimulates adenylyl cyclase indirectly, via G proteins.

TOPIC 2: THE GUSTATORY SYSTEM

KEY POINTS

✓ *What different types of taste receptor cells are found in the taste buds of the tongue?*

✓ *How do taste receptor cells transduce chemical signals?*

✓ *What parts of the brain process gustatory information?*

In mammals, the principal receptive surface for taste sensation is the tongue. Taste receptor cells are grouped together on the surface of the tongue in structures called **taste buds**, each containing about 100 receptor cells. The top surface of each taste receptor cell is covered with microvilli, which detect chemicals dissolved in the saliva on the surface of the tongue.

Four different types of taste receptor cells mediate the taste sensations: salty, sour, sweet, and bitter. Combinations of the four submodalities account for human perception of complex tastes. Salty stimuli are those with high concentrations of Na^+, whereas sour taste arises from acidic solutions. Sugars are the natural stimuli for sweet sensation, and bitter taste arises from a variety of organic compounds, including some amino acids.

Salt-sensitive receptor cells have the simplest transduction mechanism. Open Na^+ channels in the microvilli provide a path for Na^+ dissolved in the saliva to directly enter and depolarize the salt receptor cells. In sour-sensitive receptor cells, hydrogen ions (protons) directly bind to and block K^+ channels in the microvilli. The reduction in K^+ permeability then causes the receptor cell to depolarize.

Chemical substances that give rise to sweet and bitter sensations are detected by specific receptor molecules in the membrane of the microvilli. Activated receptor molecules then stimulate G proteins, in a manner similar to olfactory transduction. In sweet-sensitive taste cells, the activated G protein probably stimulates adenylyl cyclase and increases cyclic AMP. Cyclic AMP then activates protein kinase A, an enzyme that phosphorylates K^+ channels. Phosphorylated K^+ channels close, reducing K^+ permeability and depolarizing the sweet-sensitive receptor cell. In bitter-sensitive taste cells, the activated G protein stimulates phosphodiesterase, which decreases cyclic AMP and cyclic GMP inside the cell. Because the cyclic nucleotides bind to and close nonspecific cation channels in the bitter-sensitive taste cells, the resulting fall in cyclic nucleotide levels opens the channels and depolarizes the cell.

Taste receptor cells make excitatory synapses onto nerve fibers of sensory neurons, whose axons enter the brain in the **facial nerve** (cranial nerve VII) and the **glossopharyngeal nerve** (cranial nerve IX) and terminate in the **nucleus of the solitary tract** of the medulla. This nucleus also receives sensory inputs from other organs involved in homeostasis, such as the gut. The nucleus of the solitary tract sends axons to the **pontine taste nucleus** in the dorsal part of the pons and the **ventral posterior medial nucleus** of the thalamus. The pontine taste nucleus projects to the limbic system and to the ventral posterior medial nucleus in the thalamus. The thalamic nucleus relays gustatory information to the **gustatory cortex** and the **insula**. The gustatory cortex is located in the postcentral gyrus, just below and anterior to the part of somatosensory cortex that receives somatosensory inputs from the tongue. The insula is hidden from view within the lateral sulcus, a deep infolding separating the temporal lobe from the rest of the cerebral cortex.

Topic Test 2: The Gustatory System

True/False

1. Taste receptor cells synapse locally within the taste bud on sensory neurons, whose axons then project into the brain.

2. Gustatory information enters the brain by means of the facial nerve and the glossopharyngeal nerve.

3. In all types of taste receptor cell, chemotransduction involves receptor molecules that are coupled to G proteins.

4. Some taste receptor cells depolarize in response to taste stimuli, whereas others hyperpolarize in response to taste stimuli.

5. All gustatory information reaches the cerebral cortex by means of the ventral posterior medial nucleus of the thalamus.

6. The axons of sensory neurons that innervate the taste buds terminate in the pontine taste nucleus.

7. In some taste receptor cells, taste stimuli act by closing ion channels, whereas in other taste receptor cells, taste stimuli act by opening ion channels.

Multiple Choice

8. Cyclic nucleotides are thought to be involved in chemotransduction in
   a. sweet-sensitive taste receptor cells.
   b. sour-sensitive taste receptor cells.
   c. bitter-sensitive taste receptor cells.
   d. a and b but not c.
   e. a and c but not b.

9. Which of the following brain regions is part of the gustatory pathway?
   a. insula
   b. nucleus of the solitary tract
   c. limbic system
   d. all of the above
   e. none of the above

Short Answer

10. What are the four submodalities of taste?

Topic Test 2: Answers

1. **True.** Taste receptor cells make excitatory synapses onto sensory fibers that then carry gustatory information into the brain.

2. **True.** Axons of sensory neurons that receive synaptic input from taste receptor cells are found in both the facial and glossopharyngeal nerves. The facial nerve carries inputs from the anterior part of the tongue, and the glossopharyngeal nerve carries fibers from the posterior part of the tongue.

3. **False.** Receptors coupled to G proteins are involved in transduction in sweet-sensitive and bitter-sensitive taste receptor cells. However, salt-sensitive and sour-sensitive taste receptor cells operate by a direct interaction between the taste stimulus and ion channels, without acting via G proteins.

4. **False.** All taste receptor cells depolarize in response to a stimulus.

5. **True.** As with other sensory systems that reach the cerebral cortex, taste information is relayed through the thalamus before reaching the cortex. The ventral posterior medial nucleus is the thalamic nucleus for taste.

6. **False.** Inputs from taste buds first terminate in the nucleus of the solitary tract in the medulla, which then projects to the pontine taste nucleus and the thalamus.

7. **True.** For example, in bitter-sensitive taste cells, stimuli open nonspecific cation channels, but in sweet-sensitive taste cells stimuli close K^+ channels. Sour-sensitive taste cells also depolarize because K^+ channels are directly blocked by the taste stimulus (protons).

8. **e.** In sweet-sensitive and bitter-sensitive receptor cells, taste stimuli activate G protein-coupled receptors. In both cases, the enzyme stimulated by G proteins affects cyclic nucleotide levels inside the cell. For sweet stimuli, cyclic AMP is increased by the stimulus, but bitter stimuli decrease cyclic nucleotide levels.

9. **d.** The insula is part of the cerebral cortex that receives gustatory input from the thalamus. The limbic system receives inputs from the pontine taste nucleus, and the nucleus of the solitary tract receives incoming axons from sensory neurons that innervate taste buds.

10. The four submodalities of taste are salt, sour, sweet, and bitter. Individual taste receptor cells, however, are typically able to respond to more than one submodality. For example, a particular taste cell might depolarize strongly to a salty stimulus but might also depolarize weakly to sour and bitter stimuli. A different taste cell might give the largest response to sweet stimuli, but also respond less strongly to sour, bitter, and salty stimuli. Thus, although each cell has a preferred submodality, the nervous system can unequivocally identify the stimulus only by comparing the pattern of activation across all receptors.

IN THE CLINIC

In humans, the ability to detect odors varies substantially from one individual to another. Some individuals are less sensitive to odors in general, much like the reduced sensitivity that occurs in all of us when we have nasal congestion. However, in some individuals, the overall sensitivity to odors is normal but the ability to detect a particular substance or related group of substances is lacking. Such specific anosmia may result from mutations in the gene encoding a particular one of the many types of olfactory receptor molecules normally found in olfactory receptor cells. Without the receptor molecule, the presence of the odorant that binds to that particular receptor cannot be detected. Frequently, the affected person is not aware of any defect, probably because the absence of a single receptor type out of 1,000 different types does not alter olfactory perception appreciably. More generalized reduction in olfactory sensitivity (hyposmia) often occurs with aging. Because of the importance of olfaction in the "taste" of food, hyposmia frequently leads to clinical complaints.

Chapter Test

True/False

1. The chemical stimulus is directly linked to ion channels in olfactory transduction but not in taste transduction.

2. G proteins are involved in chemotransduction in sweet and bitter taste transduction but not in salt and sour taste transduction.

3. Chemical stimuli depolarize olfactory receptor cells but hyperpolarize taste receptor cells.

4. Olfactory receptor cells have axons that project directly to the brain without intervening synapses.

5. Taste receptor cells do not have axons and instead make local synapses within the taste bud onto nerve fibers of sensory neurons.

6. Ion channels that are directly gated by cyclic nucleotides are involved in sensory transduction in both olfactory receptor cells and a subset of taste receptor cells.

7. Both olfactory and gustatory information is sent to the limbic system in the brain.

8. Mitral cells and tufted cells are neurons of the olfactory bulb that receive direct excitatory synapses from olfactory receptor cells.

9. The first nucleus in the brain that receives incoming axons from taste buds is the insula.

10. The ventral posterior medial nucleus of the thalamus receives olfactory inputs from both the pontine taste nucleus and the limbic system.

Multiple Choice

11. In the olfactory bulb,
    a. there are as many glomeruli as there are types of olfactory receptor cell.
    b. only tufted cells receive synaptic inputs from olfactory receptor neurons.
    c. each olfactory receptor cell sends axon branches to all of the glomeruli.
    d. all of the above.
    e. none of the above.

12. Taste receptor cells
    a. have about 1,000 different receptor molecules, each one specific for a particular chemical stimulus.
    b. always use G protein-coupled receptors to detect chemical stimuli.
    c. have axons that project into the medulla.
    d. all of the above.
    e. none of the above.

13. Olfactory receptor cells and taste receptor cells differ in the following way:
    a. cyclic nucleotides are used as a second messenger by olfactory cells but not by taste cells.
    b. odorant molecules directly bind to ion channels in olfactory cells but not in taste cells.
    c. chemical stimuli depolarize olfactory receptor cells by opening nonspecific cation channels, but in some types of taste receptor cells, chemical stimuli depolarize the cell by closing K^+ channels.
    d. all of the above.
    e. none of the above.

14. The olfactory pathway in the brain
    a. includes a thalamocortical projection from the medial dorsal nucleus of the thalamus to the frontal lobe of the neocortex.
    b. includes a primitive form of cortex, called paleocortex, which receives olfactory information from the olfactory bulb.

c. includes the limbic system.

d. all of the above.

e. none of the above.

15. The gustatory pathway in the brain

   a. includes a direct projection from the nucleus of the solitary tract, in the medulla, to the gustatory part of the cerebral cortex.

   b. includes a thalamocortical pathway involving the ventral posterior medial nucleus of the thalamus.

   c. does not include the limbic system.

   d. all of the above.

   e. none of the above.

16. The parts of the cerebral cortex involved in chemosensory processing

   a. include a portion of the postcentral gyrus, the part of the cortex principally associated with the somatosensory system.

   b. are located exclusively in the frontal lobe and temporal lobe.

   c. receive inputs only from the paleocortex.

   d. a and c but not b.

   e. a, b, and c.

Short Answer

17. List the sequence of synaptic relays leading from the olfactory epithelium to the part of the neocortex that receives olfactory information.

18. Which cranial nerves carry chemosensory information?

Essay

19. Describe the most direct pathway leading from the tongue to the cerebral cortex.

20. Compare and contrast the sensory transduction process in olfactory and taste receptor cells.

Chapter Test Answers

1. **F**   2. **T**   3. **F**   4. **T**   5. **T**   6. **T**   7. **T**   8. **T**   9. **F**   10. **F**   11. **a**   12. **e**

13. **c**   14. **d**   15. **b**   16. **a**

17. Olfactory epithelium to olfactory bulb, which projects to the olfactory cortex (paleocortex). From the olfactory cortex to the medial dorsal nucleus of the thalamus, which projects to the olfactory neocortex, in the frontal lobes.

18. cranial nerve I (olfactory nerve), cranial nerve VII (facial nerve), cranial nerve IX (glossopharyngeal nerve).

19. From the tongue, axons of sensory neurons that innervate the taste buds enter the medulla via cranial nerves VII and IX and make synapses on neurons in the nucleus of the solitary tract, which in turn sends axons to the ventral posterior medial nucleus of

the thalamus. The thalamic neurons relay gustatory signals to the gustatory cortex, in the most lateral and anterior part of the postcentral gyrus, and to the insula.

20. In olfactory receptor cells and in sweet-sensitive and bitter-sensitive taste cells, the chemical signal is detected by receptor molecules that are linked indirectly to ion channels via G proteins. In these cells, G proteins affect the cyclic nucleotide levels. The linkage between cyclic nucleotides and ionic permeability is different in olfactory and taste cells, however. In olfactory cells, stimuli increase cyclic AMP, which directly opens nonspecific cation channels. In sweet-sensitive taste cells, stimuli increase cyclic AMP as in olfactory cells, but cyclic AMP then promotes phosphorylation of K^+ channels by protein kinase A, and the phosphorylated channels close. In bitter-sensitive taste cells, stimuli decrease cyclic nucleotide levels by activation of phosphodiesterase. In this case, however, cyclic nucleotides close nonspecific cation channels, which is opposite to the action of cyclic nucleotides in olfactory cells. In salt-sensitive and sour-sensitive taste cells, the chemical signal (sodium ions and protons, respectively) interacts directly with ion channels. Sodium ions directly pass through open Na^+ channels and depolarize salt-sensitive cells, whereas protons directly block K^+ channels in sour-sensitive cells.

Check Your Performance:

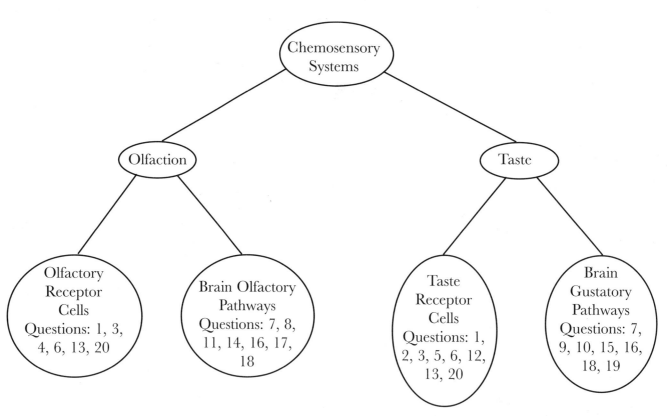

Note the number of questions in each grouping that you got wrong on the chapter test. Identify areas where you need further review and go back to relevant parts of this chapter.

Further help: If you continue to have difficulty with the concepts, a detailed presentation of these topics can be found in chapter 19 of *Neurobiology: Molecules, Cells, and Systems*, published by Blackwell Science, Inc.

First Midterm Exam

True/False

1. The central nervous system consists of the brain and spinal cord. (Chapter 1)

2. The telencephalon and diencephalon are the two main divisions of the forebrain. (Chapter 1)

3. The three main divisions of the hindbrain are the cerebellum, basal ganglia, and hypothalamus. (Chapter 1)

4. During embryogenesis, the notochord induces the formation of the neuroectoderm in the overlying ectoderm. (Chapter 2)

5. The peripheral nervous system develops from the cells of the neural tube. (Chapter 2)

6. At the equilibrium potential for a permeant ion, the flux of the ion down its concentration gradient is exactly balanced by the flux of the ion down the electrical gradient across the membrane. (Chapter 3)

7. If p_{Na} were greater than p_K, the resting membrane potential of a cell would be closer to the sodium equilibrium potential than to the potassium equilibrium potential. (Chapter 3)

8. During the undershoot of the action potential, the membrane potential moves closer to the potassium equilibrium potential than at the resting potential because sodium permeability has returned to its resting level while potassium permeability remains elevated for a brief time after the action potential. (Chapter 4)

9. During the repolarizing phase of the action potential, the two mechanisms of repolarization are the closing of potassium channel n gates and the opening of the sodium channel m gates. (Chapter 4)

10. Neurotransmitter is released from a presynaptic terminal by a process of exocytosis, when synaptic vesicles fuse with the plasma membrane of the synaptic terminal in response to calcium influx. (Chapter 5)

11. A neurotransmitter that opens chloride channels in the postsynaptic cell would produce an excitatory postsynaptic potential. (Chapter 5)

12. Nociceptors respond to painful stimuli. (Chapter 6)

13. The receptive field of a secondary sensory neuron is always the same as the receptive field of the primary sensory receptors from which the secondary neuron receives synaptic input. (Chapter 6)

14. In the somatosensory system concerned with muscle senses, the muscle spindle receptors give information about muscle length, whereas the receptors of Golgi tendon organs give information about muscle tension. (Chapter 7)

15. The main spinal cord pathway for ascending information about pain and temperature is the lateral sensory tract. (Chapter 7)

16. In the retina, both the photoreceptors and the horizontal cells are depolarized in darkness and hyperpolarize in response to illumination. (Chapter 8)

17. The neurons of the primary visual cortex have center-surround receptive fields. (Chapter 8)

18. The first auditory relay station in the brain is the cochlear nucleus, which is organized in a spatial representation of sound frequency, called a tonotopic map. (Chapter 9)

19. All hair cells of the cochlea respond best to low-frequency sounds (<1000 Hz). (Chapter 9)

20. Olfactory receptor cells have axons that project directly to the brain without intervening synapses. (Chapter 10)

21. G-proteins are involved in the linkage between the chemical stimulus and the change in ionic permeability in sweet and bitter taste transduction, but not in salt and sour taste transduction. (Chapter 10)

Multiple Choice

22. During development of the nervous system (Chapter 2)
    a. the neuroectoderm is induced during gastrulation in the part of the ectoderm overlying the notochord.
    b. the cells of the peripheral nervous system arise from the neural crest.
    c. the neural tube gives rise to the central nervous system.
    d. all of the above are correct.
    e. none of the above is correct.

23. Suppose that in a neuron, the sodium equilibrium potential is +58 mV, the potassium equilibrium potential is −80 mV, and the resting membrane potential of the neuron is −70 mV.
    a. In this neuron, the potassium permeability of the membrane is less than the sodium permeability.
    b. If the sodium permeability and the potassium permeability were both doubled, the resting membrane potential of the neuron would remain at −70 mV.
    c. In this neuron, the internal potassium concentration is less than the external potassium concentration.
    d. In this neuron, both sodium and potassium are at equilibrium at the resting membrane potential of the cell.
    e. In this neuron, the membrane is permeable to potassium but not to sodium.

24. During an action potential (Chapter 4)
    a. the membrane potential at the peak of the action potential is close to the sodium equilibrium potential.
    b. the upstroke (depolarizing phase) of the action potential is produced by a large increase in sodium permeability as voltage-dependent sodium channels open.
    c. the refractory period corresponds to the period after an action potential when voltage-dependent potassium channels remain open, moving the membrane potential nearer to the potassium equilibrium potential.
    d. a and b, but not c.
    e. a, b, and c.

25. Regarding synaptic transmission, which of the following statements is *not* correct? (Chapter 5)

a. Neurotransmitter release is triggered by calcium influx through voltage-sensitive calcium channels, which open in response to the depolarization produced by the arrival of an action potential in the synaptic terminal.

b. Neurotransmitter is released from the synaptic terminal by exocytosis, when synaptic vesicles fuse with the plasma membrane of the terminal.

c. An inhibitory neurotransmitter produces inhibition of a postsynaptic neuron by preventing excitatory neurotransmitters from binding to their receptors at excitatory synapses onto the postsynaptic neuron.

d. An excitatory postsynaptic potential could be produced by a neurotransmitter that opens nonspecific cation channels that are equally permeable to sodium and potassium ions.

e. The postsynaptic action of a neurotransmitter is mediated by neurotransmitter receptor molecules in the postsynaptic membrane, which bind the neurotransmitter and either directly or indirectly alter the ionic permeability of the postsynaptic cell.

26. In a primary sensory receptor cell (Chapter 6)
    a. the receptor potential is the change in membrane potential produced by synaptic inputs from neighboring receptor cells.
    b. sensory transduction is required for generation of a receptor potential only in exteroceptors, and not in interoceptors.
    c. lateral inhibition refers to the fact that sensory stimuli outside the receptive field inhibit the receptor cell, whereas sensory stimuli within the receptive field excite the receptor cell.
    d. receptive fields always have a center-surround organization, with opposite responses from the center and surround.
    e. the receptive field is the region of the sensory surface where a sensory stimulus affects the membrane potential of the receptor.

27. In the somatosensory system, which of the following statements is *not* correct? (Chapter 7)
    a. All the axons of primary sensory receptor cells terminate locally in the spinal cord and do not send branches into the brain.
    b. The primary somatosensory cortex is located in the postcentral gyrus of the cerebral cortex.
    c. The ascending axons of neurons that carry information about touch, pressure, vibration, and proprioception to the brain are found mostly in the dorsal columns of the spinal cord.
    d. The ascending sensory axons of the dorsal columns leave the spinal cord and terminate in the dorsal column nuclei of the medulla, the gracile nucleus, and the cuneate nucleus.
    e. The three major spinal pathways for ascending sensory information are the dorsal columns, the lateral sensory tract, and the spinocerebellar tract.

28. In photoreceptors of the retina, illumination causes the cells to hyperpolarize because (Chapter 8)
    a. rhodopsin directly binds to sodium channels in the plasma membrane and causes the channels to close.
    b. the G-protein transducin is activated when rhodopsin absorbs light, and activated transducin binds to and opens potassium channels in the plasma membrane.

c. illumination causes the level of the second messenger, cGMP, to decline inside the cell, and cGMP directly binds to and opens nonspecific cation channels in the plasma membrane.

d. all of the above are correct.

e. none of the above is correct.

29. In the auditory system (Chapter 9)

a. hair cells of the cochlea have axons that project directly to the brain in the auditory nerve.

b. the hair cells of the cochlea make synapses onto neurons of the spiral ganglion.

c. hair cells of the cochlea hyperpolarize in response to high-frequency sound stimuli and depolarize in response to low-frequency sound stimuli.

d. all of the above.

e. none of the above.

30. In the olfactory system (Chapter 10)

a. primary olfactory receptor neurons have axons that project directly to the brain, where they make synapses onto neurons of the olfactory bulb.

b. olfactory stimuli depolarize primary olfactory receptor neurons by means of G-protein-coupled receptor molecules located in the cilia of the receptor neurons.

c. each glomerulus in the olfactory bulb receives synaptic inputs from the primary olfactory receptor neurons of one particular type, among the thousand or so different types of receptor cell found in the olfactory epithelium.

d. all of the above.

e. none of the above.

Answers

1. **T** 2. **T** 3. **F** 4. **T** 5. **F** 6. **T** 7. **T** 8. **T** 9. **F** 10. **T** 11. **F** 12. **T** 13. **F** 14. **T** 15. **T** 16. **T** 17. **F** 18. **T** 19. **F** 20. **T** 21. **T** 22. **d** 23. **b** 24. **d** 25. **c** 26. **e** 27. **a** 28. **c** 29. **b** 30. **d**

UNIT IV:
MOTOR CONTROL SYSTEMS

CHAPTER 11

Neural Control of Muscle Contraction

Motor neurons are the final output of the nervous system, and their most readily apparent target is the skeletal musculature. This chapter describes the control of muscles by motor neurons, beginning with the molecular mechanisms of contraction and then turning to neural mechanisms that produce smooth and graded contractions of muscles.

ESSENTIAL BACKGROUND

- Gross anatomy of muscles
- Ionic permeability and membrane potential
- Action potential
- ATP as a cellular energy source

TOPIC 1: EXCITATION-CONTRACTION COUPLING

KEY POINTS

✓ *What is the anatomical organization of muscle cells?*

✓ *What protein molecules make up the contractile apparatus of muscle cells?*

✓ *How is muscle contraction triggered and terminated?*

Skeletal muscles consist of striated muscle cells (muscle fibers), which are bundles of smaller fibers called **myofibrils**. Myofibrils have a repeating pattern of crosswise stripes (striations), called the A band, I band, M line, and Z line. The I band is a light region with a dark Z line at its center, and the A band is a darker region separating two successive I bands, with the darker M line at its center. The repeating unit of striation is the **sarcomere**, which extends from one Z line to the next. The striations arise from **thick filaments** and **thin filaments**, which are arranged in parallel (**Figure 11.1A**). Groups of thin filaments are connected at their midlines by the Z line, and thick filaments are connected at the M line. The thick and thin filaments partially overlap, and the I band corresponds to the region where only thin filaments are found.

Thick filaments have appendages that extend to neighboring thin filaments in the region of overlap, forming **cross bridges** between the filaments. Thick filaments are aggregates of **myosin**, a protein that has a fibrous tail and a globular head. Aggregated tails of myosin mole-

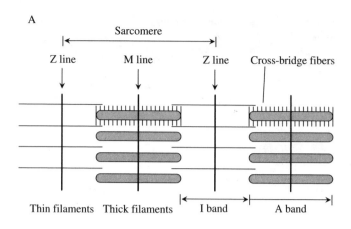

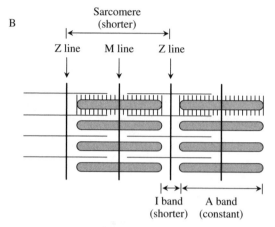

Figure 11.1 Schematic representation of the relationships between thick and thin filaments of a myofibril in (A) a relaxed muscle and (B) a contracted muscle.

cules form the backbone of thick filaments, and the globular heads form the cross bridges. Thin filaments consist of the protein **actin**, together with the associated proteins **troponin** and **tropomyosin**.

When a muscle cell contracts, each sarcomere shortens. The width of the A band remains constant during contraction, and the shortening of the sarcomere results from the I band becoming narrower during contraction (**Figure 11.1B**). Neither thick nor thin filaments change in length during contraction. Instead, shortening occurs because the filaments slide past each other, increasing the region of overlap (Figure 11.1B). Sliding of filaments during contraction results from the interaction between myosin and actin. Each actin molecule has a binding site for the globular head of myosin. Myosin is an ATPase that hydrolyzes ATP and stores the released energy. When myosin binds to actin, the stored energy is released and the globular head of myosin moves with respect to the fibrous tail. Because the globular head is attached to the thin filament, movement of the head causes the thin filament to move relative to the thick filament.

In resting muscle, the interaction between myosin and actin is prevented by tropomyosin and troponin. Tropomyosin blocks the myosin binding site on actin, and troponin locks tropomyosin in the blocking position. To trigger contraction, calcium ions bind to troponin, which alters the interaction between troponin and tropomyosin. Tropomyosin then moves and uncovers the myosin binding site, allowing cross bridge interaction and filament sliding. Calcium ions to trigger contraction come from a special intracellular compartment called the **sarcoplasmic**

reticulum. The sarcoplasmic reticulum is a membrane-bound storehouse for Ca^{2+} inside the cell, surrounding the myofibrils. Calcium is released from the sarcoplasmic reticulum in response to an action potential in the muscle cell. Depolarization produced by an action potential spreads to the interior of the muscle cell along invaginations of the plasma membrane, called **transverse tubules**. Depolarization of transverse tubules activates Ca^{2+} channels in the sarcoplasmic reticulum, and Ca^{2+} moves out of the sarcoplasmic reticulum, binds to troponin, and initiates contraction.

Topic Test 1: Excitation-Contraction Coupling

True/False

1. A sarcomere is defined as extending from one M line to the next M line along the myofibril.

2. Upon contraction, each sarcomere remains constant in length, but the width of the I band becomes shorter.

3. During contraction, neither thick nor thin filaments change in length, but the degree of overlap of the filaments increases.

4. Contraction is triggered when Ca^{2+} binds to troponin, which is found on thin filaments.

5. The sarcoplasmic reticulum is formed by infoldings of the plasma membrane and allows depolarization during an action potential to spread to the interior of the muscle cell.

6. Globular heads of actin form the cross bridges that connect the thick and thin filaments.

7. Thick filaments and thin filaments slide past each other during contraction because of a conformation change in myosin.

Multiple Choice

8. ATP is required in muscle contraction to
   a. provide energy to fuel the sliding of filaments past each other.
   b. break the bond between myosin and actin after myosin has released its stored energy, so that the cross bridge cycle can continue to produce further sliding.
   c. terminate the contraction by fueling Ca^{2+} pumps that transport released Ca^{2+} back into the sarcoplasmic reticulum.
   d. all of the above are correct.

9. According to the sliding filament mechanism of muscle contraction,
   a. the width of the A band does not change during contraction because the A band corresponds to the length of thick filaments, which does not change during contraction.
   b. the width of the I band does not change during contraction because the I band corresponds to the length of thin filaments, which does not change during contraction.
   c. myofibrils of a muscle cell slide past each other during contraction.
   d. a and b but not c.
   e. a and c but not b.

10. Contraction is initiated when
    a. an action potential is triggered in the muscle cell.
    b. calcium ions are released from the sarcoplasmic reticulum.
    c. calcium ions bind to the calcium-binding protein of the thin filaments, troponin.
    d. tropomyosin alters its position on the thin filament, uncovering the myosin binding site of actin.
    e. all of the above.

Topic Test 1: Answers

1. **False.** The sarcomere is the basic unit of striation, but the sarcomere is defined as extending from one Z line to the next Z line. The M line is located at the midpoint of each sarcomere.

2. **False.** Upon contraction, the I band does shorten. However, each sarcomere must also shorten, as the distance between successive Z lines decreases.

3. **True.** Thick and thin filaments slide past each other during contraction, but the lengths of the filaments do not change. As thin filaments are pulled past thick filaments, the region of overlap becomes larger and the region where thin filaments do not overlap thick filaments (the I band) becomes shorter.

4. **True.** Troponin is a calcium-binding protein associated with the thin filaments. Without Ca^{2+}, troponin keeps tropomyosin in a position to block access to the myosin binding site of actin. When Ca^{2+} binds to troponin, tropomyosin reveals the myosin binding site, and interaction between myosin and actin is allowed.

5. **False.** The transverse tubules are actually the infoldings of the plasma membrane that carry depolarization to the interior of the muscle cell. The sarcoplasmic reticulum is an intracellular storehouse for Ca^{2+}, surrounding the myofibrils. The membrane of the sarcoplasmic reticulum is entirely separate from the plasma membrane.

6. **False.** Cross bridges between thick and thin filaments are actually globular heads of myosin molecules.

7. **True.** The globular head of myosin hydrolyzes ATP, and the released chemical energy causes a conformation change in the globular head of myosin, shifting the geometrical relationship between the globular head and fibrous tail of the molecule. During contraction, "energized" myosin binds to the myosin binding site of actin on a neighboring thin filament. When energized myosin binds to actin, the stored energy is released, and the globular head relaxes to its alternate conformation. Because the reverse conformation change produces movement of the head with respect to the tail, the thin filament attached to the head also moves with respect to the thick filament.

8. **d.** ATP has multiple roles in muscle contraction. First, hydrolysis of ATP provides energy to drive filament sliding (statement a). After the cross bridge movement has occurred, the bond between actin and the relaxed myosin must be broken to allow the movement cycle to repeat. The unbinding of myosin from actin is stimulated by binding of a new molecule of ATP to myosin (statement b). If ATP is not available, the cross bridge attachment cannot be broken, and the muscle becomes rigid. This occurs during rigor mortis. Another role for ATP is termination of contraction (statement c). After a

contraction, the resting state must be restored again. To terminate contraction, Ca^{2+} released from the sarcoplasmic reticulum must be removed from the cytoplasm. This is accomplished by Ca^{2+} pumps in the membrane of the sarcoplasmic reticulum. The Ca^{2+} pump uses energy from hydrolysis of ATP to drive uptake of Ca^{2+} into the sarcoplasmic reticulum.

9. **a.** The A band remains constant during contraction, because the A band corresponds to the length of the thick filaments, which do not change in length. Statement b is incorrect because the I band in fact shortens during contraction. The I band is the region where thin filaments do not overlap thick filaments. Because overlap between thin and thick filaments increases during contraction, the region of nonoverlap necessarily decreases. Statement c is incorrect because the filaments within a myofibril slide past each other during contractions, but the myofibrils making up the muscle cell do not slide past each other.

10. **e.** Action potentials trigger contraction of muscle cells, producing depolarization that spreads to the cell interior along the transverse tubules, which are infoldings of the plasma membrane. Depolarization of transverse tubules triggers release of Ca^{2+} from the sarcoplasmic reticulum, and Ca^{2+} then binds to troponin on the thin filaments. Binding of Ca^{2+} to troponin allows tropomyosin to reveal the binding site on actin for attachment of the myosin cross bridge.

TOPIC 2: NEURAL CONTROL OF MUSCLE CONTRACTION

KEY POINTS

✓ *How are twitches of single muscle cells integrated to produce smooth movements?*

✓ *What is a motor unit?*

✓ *How does the nervous system avoid muscle fatigue?*

A **motor unit** consists of a single motor neuron and all its synaptically connected muscle cells. When the motor neuron fires an action potential, all muscle fibers in the motor unit twitch together. The strength of contraction generated by a motor unit depends on the number of muscle fibers in that motor unit, which varies considerably among the motor units making up a muscle. The tension developed by a muscle depends on the number of active motor units and the frequency of contraction of each motor unit. Increasing muscle tension by increasing the number of active motor neurons is called motor neuron **recruitment**. When muscle tension is low, small motor units are recruited first, ensuring that added increments of tension are small and preventing large jerky increases in tension. As tension rises, further increases in tension must be larger to make a significant difference. Thus, larger motor units are added, resulting in larger increments of tension when the background tension is already high. Recruitment according to motor unit size is called the **size principle**.

The tension in a muscle cell lasts for tens to hundreds of milliseconds after an action potential, and so tension summates during a series of action potentials at a rapid rate. When the frequency of action potentials in a motor unit increases, muscle tension climbs until it reaches a steady plateau state, called **tetanus**. Muscles produce different types of contraction, depending on the load the muscle is working against. If the muscle cannot lift the load, muscle length does not change, and the contraction is an **isometric contraction**. If the muscle lifts the load, the muscle shortens, tension in the muscle remains constant, and the contraction is an **isotonic contraction**.

Normally, all motor neurons of a muscle are not active simultaneously during a maintained contraction. Instead, activity of individual motor neurons takes place in periodic bursts that occur asynchronously in different motor neurons. This process reduces fatigue by allowing individual motor units to rest periodically during a maintained contraction.

Topic Test 2: Neural Control of Muscle Contraction

True/False

1. A motor unit consists of a single motor neuron and all the muscle fibers that receive synapses from that motor neuron.

2. As tension in a muscle increases, the first motor units to be recruited are the largest ones, with the largest number of muscle cells.

3. During an isometric contraction, the length of the muscle does not change because the force exerted by the muscle is insufficient to move the load.

4. In a skeletal muscle cell, a contraction lasts only as long as the depolarization produced by the muscle action potential.

5. During a maintained contraction, all motor neurons supplying the muscle fire action potentials continually at a high rate.

6. The size principle states that all motor units are the same size (i.e., all have the same number of muscle cells).

7. Muscle tension increases during an isotonic contraction but does not change during an isometric contraction.

Multiple Choice

8. Gradation in the overall strength of muscle contraction is achieved by
   a. varying the total number of motor neurons activated, and thus the total number of motor units contracting.
   b. varying the frequency of action potentials in the motor neuron of a single motor unit.
   c. varying the number of muscle cells that contract in a single motor unit.
   d. a and b but not c.
   e. a and c but not b.

Topic Test 2: Answers

1. **True.** In vertebrate muscles, each skeletal muscle cell receives synaptic input from only one motor neuron, but each motor neuron contacts several muscle cells. Each action potential in a motor neuron triggers an action potential in every muscle cell contacted by the motor neuron, and so all the muscle cells of the motor unit contract together when the motor neuron fires. Thus, the motor unit is an irreducible unit of contraction.

2. **False.** The first motor units to be activated in a relaxed muscle are the smallest motor units, containing the smallest number of muscle fibers. As the tension grows, further increases in tension are achieved by activating progressively larger motor units.

3. **True.** When activated, a muscle can actually shorten only if the tension is sufficient to move the load attached to the muscle. If the load is too great, shortening will not occur, but muscle tension will still increase nevertheless. Such an attempt to contract, without shortening, is called an isometric ("same length") contraction. If tension is sufficient to move the load, the contraction is called isotonic ("same tension") because tension remains constant once it is sufficient to move the load.

4. **False.** Action potentials trigger contraction in a skeletal muscle cell, but duration of contraction is governed by the contractile apparatus and the mechanical properties of the muscle cell. Muscle cells fall into two classes based on duration of contraction: fast and slow muscle fibers. In fast fibers, tension reaches a peak within 10 to 50 milliseconds after an action potential and returns to rest within about 100 milliseconds. In slow fibers, peak tension may occur 200 milliseconds after an action potential, and return to rest is correspondingly slower.

5. **False.** During maintained contraction, motor neurons fire in bursts, separated by quiet periods. The quiet periods avoid fatigue in the muscle cells of the motor unit. Bursts in different motor neurons are staggered in time, so that different motor units fire and rest at different times.

6. **False.** The size principle refers to the order of recruitment of motor units as tension increases in a muscle.

7. **False.** Tension increases in both isotonic and isometric contractions. The difference is that in an isotonic contraction, the tension becomes constant once it is sufficient to lift the load.

8. **d.** Recruitment of different motor units is one means to control tension in a muscle. Activation of more motor neurons produces larger tension. Also, within broad limits, the higher the frequency of action potentials in the motor neurons, the greater the tension in the muscle. Thus, statements a and b are both correct. Statement c is false, because each motor unit has a fixed size, determined by the number of muscle cells that receive synaptic contacts from the motor neuron.

IN THE CLINIC

Many human diseases affect the motor unit. Proper functioning of the motor unit requires four transmission stages: the cell body of the motor neuron in the spinal cord, the axon of the motor neuron in the peripheral nerve, the neuromuscular junction, and the muscle cells. Diseases of the motor unit are distinguished based on which of the four stages is affected. Diseases that affect the muscle are called myopathies, whereas neuropathy refers to diseases that affect the motor neuron or the peripheral nerve. Diseases of the neuromuscular junction are a separate category. Because neurogenic diseases often lead to atrophy of the muscle, the distinction between neuropathy and myopathy is often blurred. Perhaps the best-known disease that affects the motor neurons is amyotrophic lateral sclerosis (Lou Gehrig's disease), which is an inherited disorder in which motor neurons die. The genetic defect has been localized to a gene encoding an enzyme involved in removal of toxic free radicals, which are a byproduct of oxidation reactions. It is not yet clear why abnormal handling of free radicals preferentially affects motor neurons.

Neuropathies of the peripheral nerve often interfere with motor unit function by blocking action potential conduction in the nerve. An example is Guillain-Barré syndrome, which is thought to be an autoimmune disease in which antibodies attack and destroy the myelin sheath of axons in peripheral nerves. Duchenne muscular dystrophy is an inherited disorder that leads to progressive degeneration of skeletal muscle cells. The underlying defect is a mutation in the gene for a muscle protein called dystrophin, which is involved in linking plasma membrane proteins to the cytoskeleton. Diseases that block neuromuscular synaptic transmission also disrupt the function of motor units. Two examples of auto-immune diseases that target the neuromuscular junction are myasthenia gravis, in which antibodies attack acetylcholine receptors, and Lambert-Eaton myasthenic syndrome, in which antibodies attack voltage-dependent calcium channels of the motor nerve synaptic terminal.

Chapter Test

True/False

1. A single muscle cell consists of a bundle of smaller elements called myofibrils.
2. During contraction, there is no change in length of each sarcomere of a myofibril.
3. Movements of the cross bridges connecting thick and thin filaments propel the sliding of filaments past each other during contraction of a muscle cell.
4. Depolarization of the membrane of the sarcoplasmic reticulum during an action potential is the trigger for contraction of a muscle cell.
5. Binding of Ca^{2+} to tropomyosin unblocks the myosin binding sites of actin to initiate contraction.
6. Energy to fuel muscle contraction is provided by ATP hydrolysis by myosin.
7. Calcium ions are released from the sarcoplasmic reticulum to trigger contraction and pumped back into the sarcoplasmic reticulum to terminate contraction.
8. Asynchronous activation of motor units avoids muscle fatigue during prolonged contraction.
9. If a motor neuron fires a burst of five action potentials within 0.25 seconds, the tension achieved in the muscle innervated by the motor neuron would be greater than if the neuron fires a burst of five action potentials spread out over 1 second.
10. All motor units in a muscle have the same number of muscle cells.

Multiple Choice

11. The sarcomere
    a. extends from one Z line to the next Z line in the repeating striations of a myofibril.
    b. corresponds to the region occupied by thick filaments in each repeating striped pattern in a myofibril.

c. does not change in length during a contraction, because neither thick nor thin filaments change in length.

d. all of the above.

e. none of the above.

12. Myosin

a. has a fibrous tail region, where ATP binds and is hydrolyzed.

b. is an ATPase, in which ATP hydrolysis alters the spatial relation of the globular head and fibrous tail.

c. binds to actin in the energized state but unbinds immediately from actin when the energy is released.

d. all of the above.

e. none of the above.

13. In the muscle cell, calcium ions

a. are stored in an intracellular compartment called the sarcoplasmic reticulum.

b. are released into the cytoplasm in response to depolarization of the transverse tubules.

c. are removed from the cytoplasm by ATP-driven calcium pumps to terminate contraction.

d. all of the above.

e. none of the above.

14. The strength of contraction produced by action potentials in a single motor neuron depends on

a. the number of muscle cells that receive synapses from the motor neuron.

b. the frequency of action potentials in the motor neuron.

c. whether the motor unit has fired at a high frequency for some time or has recently rested.

d. all of the above.

e. none of the above.

15. Motor unit A consists of 10 muscle cells, motor unit B consists of 30 muscle cells, and motor unit C consists of 90 muscle cells. In what order would these three motor units be recruited during increasing tension?

a. A, then B, then C.

b. C, then B, then A.

c. C, then A, then B.

d. A, B, and C would all be stimulated at the same time.

e. None of the above.

16. During contraction of skeletal muscle,

a. calcium ions to trigger contraction enter the muscle cell from the extracellular fluid.

b. all motor units of the muscle will always be active at the same time.

c. each muscle cell contracts and relaxes again within about 10 milliseconds.

d. all of the above.

e. none of the above.

Short Answer

17. Describe why the A band remains constant but the I band becomes shorter during contraction of a muscle cell.

18. Identify the following: tropomyosin, troponin, myosin, actin.

19. Briefly describe the size principle and state why it is functionally significant for generating smoothly graded contraction of skeletal muscles.

Essay

20. List the sequence of events between generation of an action potential in a muscle cell and the contraction and subsequent relaxation of the cell.

Chapter Test Answers

1. **T** 2. **F** 3. **T** 4. **F** 5. **F** 6. **T** 7. **T** 8. **T** 9. **T** 10. **F** 11. **a** 12. **b**

13. **d** 14. **d** 15. **a** 16. **e**

17. The A band remains constant during contraction because the A band corresponds to the thick filaments, which do not change in length during contraction. The I band is the region where thin filaments do not overlap thick filaments. As thin filaments slide past the thick filaments during contraction, this region of nonoverlap decreases and so the I band becomes narrower.

18. Tropomyosin: a regulatory protein on the thin filament that controls access to the myosin binding site of actin. Troponin: a calcium-binding protein on the thin filament that controls the position of tropomyosin. Myosin: the thick filaments are aggregates of myosin, which forms cross bridges with the thin filaments and uses energy from hydrolysis of ATP to propel filament movement. Actin: a myosin-binding protein that makes up the backbone of the thin filaments.

19. The size principle states that as muscle tension progressively increases, motor neurons are recruited in order of the size of their motor units. Smaller motor units produce small increments in muscle tension, whereas larger motor units produce larger increments. When the load tension is low, adding tension by selectively activating small motor units ensures smooth contraction. When tension is high, larger motor units must be activated to significantly increase tension.

20. 1) The action potential propagates along the muscle fiber, producing a large depolarization of the plasma membrane. 2) Depolarization produced by the action potential spreads to the interior of the fiber along the transverse tubules. 3) Depolarization of the transverse tubules causes release of Ca^{2+} by the sarcoplasmic reticulum. 4) Released calcium ions bind to troponin on the thin filaments. 5) When calcium binds to troponin, tropomyosin uncovers the myosin binding site of actin on the thin filament. 6) Globular heads of myosin molecules, which have been energized by hydrolysis of ATP, attach to actin at the myosin binding site. 7) Stored energy of activated myosin is released, propelling the thick and thin filaments past each other. Spent ADP is then released from myosin. 8) A new ATP molecule binds to myosin, breaking its attachment to the actin molecule. 9) The new ATP is hydrolyzed to re-energize myosin and return the contraction cycle to step 6. 10) Contraction continues as long as internal Ca^{2+} remains elevated. Calcium concentration declines as Ca^{2+} is pumped back into the sarcoplasmic reticulum by an ATP-driven Ca^{2+} pump.

Check Your Performance:

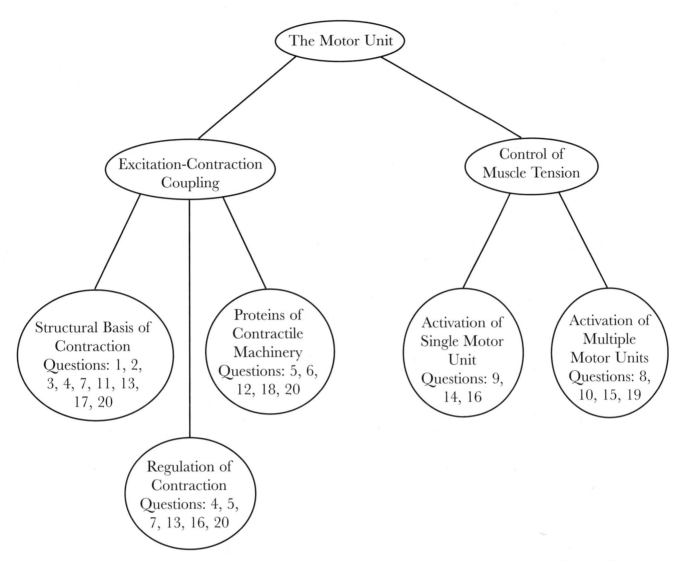

Note the number of questions in each grouping that you got wrong on the chapter test. Identify areas where you need further review and go back to relevant parts of this chapter.

Further help: If you continue to have difficulty with the concepts, a detailed presentation of these topics can be found in chapter 9 of *Neurobiology: Molecules, Cells, and Systems*, published by Blackwell Science, Inc.

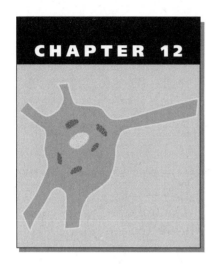

CHAPTER 12

Spinal Cord Motor Mechanisms

The motor system is organized hierarchically. Spinal levels are organized into functional circuits that carry out specific tasks, and higher levels select among spinal circuits to program more complex tasks. This chapter examines the organization of spinal cord circuits that automate commonly needed motor functions.

ESSENTIAL BACKGROUND

- **Anatomy of the spinal cord**
- **Excitatory and inhibitory synaptic transmission**
- **Muscle and skin sensory receptors (Chapter 7)**

TOPIC 1: REFLEXES CONTROLLING MUSCLE LENGTH, MUSCLE TENSION, AND LIMB POSITION

KEY POINTS

✓ *What synaptic connections underlie the myotatic reflex?*

✓ *What synaptic connections underlie the inverse myotatic reflex?*

✓ *How does the spinal cord coordinate the withdrawal reflex?*

The stretch reflex, or **myotatic reflex**, is a simple reflex activated by muscle spindle receptors (Chapter 7). Muscle stretch stimulates the sensory neurons that innervate intrafusal muscle fibers within **muscle spindles**. Axons of the sensory neurons enter the spinal cord and make excitatory synapses onto motor neurons of the same muscle and synergistic (agonistic) muscles of the same joint and onto inhibitory neurons that in turn make inhibitory synapses with motor neurons of antagonistic muscles of the same joint. This combination of synapses tends to maintain a constant muscle length—and hence a fixed joint position—by stimulating muscles that restore joint position while relaxing muscles that oppose the desired movement.

Skeletal muscles also have sensory neurons that originate from **Golgi tendon organs** (Chapter 7), which provide information about muscle tension during contraction. Axons of Golgi tendon-organ sensory neurons make excitatory synapses onto inhibitory and excitatory interneurons in the spinal cord. The inhibitory interneurons make inhibitory synapses with motor neurons of the same muscle from which the tendon-organ input originated. The excitatory interneurons make synapses with motor neurons controlling antagonist muscles of the same joint. These connections inhibit the same muscle and excite the antagonist muscles, which is the opposite of the myotatic reflex. For this reason, the tendon-organ reflex is called the **inverse myotatic reflex**.

The **withdrawal reflex** is stimulated by noxious stimuli applied to a limb (e.g., stepping on a sharp object) and involves withdrawal of the stimulated limb combined with extension of the contralateral limb. Nociceptor axons make excitatory synapses onto two sets of spinal interneurons. One set excites neurons that make excitatory synapses onto flexor motor neurons of the stimulated limb. Another set excites inhibitory interneurons, which inhibit extensor motor neurons of the stimulated limb. This combination flexes the injured limb to withdraw it from the noxious stimulus. The nociceptors also excite a third set of interneurons whose axons cross the midline and terminate in the contralateral spinal cord, where they make excitatory synapses onto excitatory and inhibitory interneurons. These interneurons excite the extensor motor neurons and inhibit the flexor motor neurons of the contralateral limb, which extends the contralateral limb so that it can support the body as the injured limb is withdrawn.

Topic Test 1: Reflexes Controlling Muscle Length, Muscle Tension, and Limb Position

True/False

1. In the myotatic reflex, sensory neurons from muscle spindles make direct inhibitory synapses onto motor neurons of antagonistic muscles.

2. The myotatic reflex is activated by passive stretch of a muscle but not by active contraction, whereas the inverse myotatic reflex is activated by active contraction but not by passive stretch.

3. The myotatic reflex is activated by sensory neurons of Golgi tendon organs.

4. Sensory neurons of a tendon organ make direct excitatory synapses onto motor neurons of the muscle attached to that tendon.

5. The withdrawal reflex involves a minimum of two synaptic relays between sensory neurons and motor neurons.

6. In the withdrawal reflex, motor neurons of muscles that extend the contralateral limb are excited, whereas the motor neurons of muscles that extend the ipsilateral (injured) limb are inhibited.

7. In spinal reflexes, the same groups of excitatory and inhibitory interneurons are used to control motor neurons in more than one type of reflex.

Multiple Choice

8. During active contraction of a muscle, stretch-sensitive sensory neurons of muscle spindles
   a. lose the ability to respond to passive stretch of the muscle.
   b. fire action potentials at a rapid rate.
   c. are still able to respond to passive stretch of the muscle.
   d. a and b but not c.
   e. b and c but not a.

9. Which of the following describes the roles of the myotatic and inverse myotatic reflexes?
   a. The myotatic reflex maintains constant muscle tension, and the inverse myotatic reflex maintains constant muscle length.
   b. The myotatic reflex controls the length of extensor muscles, and the inverse myotatic reflex controls the length of flexor muscles.
   c. The myotatic reflex flexes the ipsilateral limb and extends the contralateral limb, and the inverse myotatic reflex flexes the contralateral limb and extends the ipsilateral limb.
   d. The myotatic reflex maintains constant muscle length, and the inverse myotatic reflex maintains constant muscle tension.
   e. None of the above.

10. The neural circuitry of the inverse myotatic reflex and the withdrawal reflex differ in which of the following?
    a. The inverse myotatic reflex involves a direct synapse between sensory neurons and motor neurons, but the withdrawal reflex does not.
    b. The withdrawal reflex involves inhibitory interneurons, but the inverse myotatic reflex does not.
    c. The withdrawal reflex involves motor neurons on both sides of the spinal cord, but the inverse myotatic reflex involves motor neurons only on the ipsilateral spinal cord.
    d. The inverse myotatic reflex produces opposing effects on flexor and extensor muscles of a limb, but the withdrawal reflex excites both flexor and extensor muscles of a limb.
    e. All the above.

Topic Test 1: Answers

1. **False.** Muscle spindle receptors do lead to inhibition of antagonistic motor neurons, but the inhibition is not mediated by muscle spindle sensory neurons themselves. Instead, the sensory neurons make excitatory synapses onto interneurons, which in turn make inhibitory synapses onto antagonistic motor neurons.

2. **True.** When muscle fibers are passively stretched, little force is applied to the tendon because of elasticity of the muscle fibers. Thus, passive stretch activates muscle spindle receptors, which initiate the myotatic reflex, but not tendon organ receptors, which initiate the inverse myotatic reflex. During active contraction, force generated by the muscle is transmitted through the tendon, which deforms the tendon organ and activates the tendon sensory neurons. Active contraction does not stretch the muscle spindles, however.

3. **False.** The Golgi tendon organ activates the inverse myotatic reflex. Muscle spindle receptors activate the myotatic reflex.

4. **False.** Tendon organ sensory neurons do not make direct synapses with motor neurons, unlike muscle spindle sensory neurons. Also, activation of receptors of a particular tendon organ actually inhibits motor neurons of the same muscle. This inhibition is mediated by inhibitory interneurons excited by the tendon-organ sensory neurons.

5. **True.** Nociceptor sensory neurons synapse onto spinal interneurons that send inputs to a second group of interneurons. This second group makes synapses onto motor neurons of the ipsilateral (stimulated) limb.

6. **True.** The withdrawal reflex involves excitation and inhibition of motor neurons on both sides of the spinal cord, controlling both ipsilateral and contralateral limbs. Nociceptors activate four groups of excitatory interneurons in the spinal cord. One group activates interneurons that excite flexor motor neurons of the injured limb. Another group activates interneurons that inhibit extensor motor neurons of the injured limb. These two groups withdraw the injured limb. The other groups of interneurons contacted by nociceptors excite interneurons whose axons cross the midline and activate interneurons in the contralateral spinal cord. The contralateral interneurons excite extensor motor neurons and inhibit flexor motor neurons of the contralateral (uninjured) limb. This extends the uninjured limb to take the body weight as the injured limb withdraws.

7. **True.** Interneurons controlling motor neurons may be part of various reflexes. For instance, inhibitory interneurons controlling extensor motor neurons are excited by stretch receptors of flexor muscles of the same joint during the myotatic reflex and also by sensory neurons of tendon organs from the extensor muscle during the inverse myotatic reflex. In addition, these same extensor-muscle inhibitory interneurons are excited during the withdrawal reflex for the limb. Thus, groups of spinal excitatory and inhibitory interneurons can be thought of as modular building blocks that are used in various functional circuits.

8. **c.** When a muscle contracts, a mismatch would occur between muscle length and length of muscle spindles unless spindle length is adjusted accordingly. If this occurred, passive stretch of the muscle during an active contraction would not be detected. Thus, statement a might seem to be correct. However, during active contraction, the intrafusal muscle fibers of muscle spindles are stimulated to contract by γ motor neurons, so that the lengths of intrafusal and extrafusal muscle fibers remain matched. Thus, muscle spindles can respond to passive stretch even during active contraction.

9. **d.** The myotatic reflex is stimulated by stretch of a muscle beyond its current length. Because the stretch-sensitive sensory neurons excite motor neurons of the same muscle, the reflex opposes stretch and maintains constant muscle length. Sensory neurons of the tendon organ, however, monitor muscle tension. If tension rises above the set point for the muscle, tendon-organ sensory neurons activate the inverse myotatic reflex, leading to inhibition of motor neurons of the muscle and reduced tension. Thus, the inverse myotatic reflex maintains constant muscle tension.

10. **c.** The inverse myotatic reflex affects agonist and antagonistic muscles of a particular joint but does not affect muscles of the contralateral limb. By contrast, the withdrawal reflex includes neurons on both sides of the spinal cord, producing opposite effects on the limbs on each side.

TOPIC 2: SPINAL CIRCUITS CONTROLLING LOCOMOTION

KEY POINTS

✓ *What is a central pattern generator?*
✓ *What synaptic connections produce alternating movements on the two sides of the body during locomotion?*

Groups of spinal neurons are also used in circuits controlling locomotion. If descending commands from the brain are eliminated by transecting the spinal cord, patterns of limb movements like those used in voluntary locomotion can still be elicited. Even if sensory inputs from the limbs are eliminated, alternating stepping movements can still be observed in animals with spinal cord transections. Thus, the basic pattern of locomotor output is produced by spinal **central pattern generators** that do not require sensory input or input from the brain to generate the motor pattern.

Figure 12.1 shows synaptic connections underlying swimming in the lamprey, a primitive fish. Swimming is initiated by action potentials in command axons descending along both sides of the spinal cord from command centers in the brainstem. Command axons make excitatory connections with pattern generator neurons on both sides of the spinal cord. The pattern of action potentials in the command axons does not control the pattern of output from the motor neurons. Command axons make excitatory synapses onto four neuron types on each side of the spinal cord: motor neurons and three interneurons. A crossed interneuron makes inhibitory synapses onto neurons of the contralateral swimming circuit. An excitatory interneuron makes excitatory synapses onto neurons of the ipsilateral circuit, and an inhibitory interneuron inhibits crossed interneurons on the same side of the spinal cord.

During swimming, command axons on both sides fire action potentials steadily, providing continuous excitation to the bilateral pattern generators. When one circuit becomes active, the crossed interneuron on that side prevents neurons of the contralateral circuit from firing. As activity on the excited side proceeds, firing of the crossed interneuron declines because of reduced excitation from the ipsilateral excitatory interneuron and because of delayed inhibition from the ipsilateral inhibitory interneuron. Reduced excitation from the excitatory interneuron occurs because of Ca^{2+} influx through voltage-sensitive calcium channels. Increased internal Ca^{2+} opens Ca^{2+}-activated K^+ channels, which hyperpolarize the excitatory interneuron and terminate its activity. As activity of the crossed interneuron declines, its inhibitory influence declines and activity begins in the contralateral circuit. When the contralateral circuit fires, activation of the crossed interneuron from that side inhibits the opposite network and ensures termination of the burst on that side. In this manner, bursts of action potential alternate in the two circuits, producing alternating contractions of the body muscles required for swimming movements.

Topic Test 2: Neural Control of Muscle Contraction

True/False

1. Sensory information plays no role in alternating limb movements during quadruped locomotion.

2. Severing the connection between the brain and spinal cord abolishes all locomotor movements.

3. The spinal central pattern generator for swimming contains both excitatory and inhibitory interneurons.

4. The crossed interneuron in the central pattern generator for swimming is an inhibitory neuron that makes inhibitory synapses on all neurons of the generator circuit on the contralateral side.

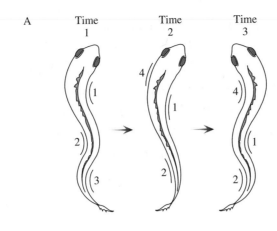

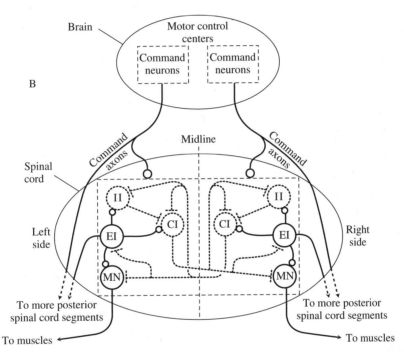

Figure 12.1 Neuronal circuitry that produces alternating contractions on the two sides of the body during swimming in the lamprey. The circuitry for a single spinal-cord segment is shown, but the same pattern of connections is repeated all along the spinal cord. (A) A schematic diagram of the undulating motion during swimming in the lamprey. At time 1, the right-side muscles are contracting, producing bending to the right, at positions 1 and 3, whereas in the middle of the animal, the contraction is on the left side, at position 2. At a somewhat time (time 2), the contraction of the right side has moved to a more posterior position (position 1), as has the contraction on the left side (position 2). A new wave of contraction on the left side has just begun (position 4). Still later (time 3), the waves of contraction on the two sides of the body have moved to still more posterior positions. (B) Circuit diagram for the central pattern generator of swimming at a single spinal-cord segment. The circuits on the two sides of the spinal cord (enclosed in dashed boxes) are mirror images of each other. The network is activated by descending excitatory inputs arising from swimming command centers in the brain. Inhibitory neurons are indicated with dashed lines. Abbreviations: MN, motoneuron; EI, excitatory interneuron; II, inhibitory interneuron; CI, crossed interneuron.

5. The alternating pattern of activity in the central pattern generators on the two sides of the spinal cord is explained entirely by the pattern of synaptic connections.

6. Although the central pattern generators for swimming in the lamprey can produce the pattern of motor output without sensory information, sensory feedback reinforces and adjusts the timing of alternating bursts of activity on the two sides of the spinal cord.

7. Groups of excitatory and inhibitory interneurons used for reflexes, such as the withdrawal reflex, do not overlap with groups of interneurons used during locomotion.

Multiple Choice

8. In the central pattern generator for swimming, alternating activity of motor neurons on each side of the spinal cord results from
   a. changes in activity of the crossed interneurons.
   b. declining activity of excitatory interneurons during a burst.
   c. inhibition by inhibitory interneurons on each side of the spinal cord.
   d. a and b but not c.
   e. a, b, and c.

Topic Test 2: Answers

1. **False.** Although central pattern generators in the spinal cord can produce the basic stepping movements of locomotion, the movements are stereotyped and easily disrupted without sensory information from the limbs. Information about joint position and muscle tension is normally used during walking to adapt the motor pattern to changing conditions. Thus, both automatic circuits and sensory information are required to produce normal locomotion.

2. **False.** The basic pattern of locomotion is produced by neural circuits in the spinal cord. If the circuits are stimulated, they are capable of producing movements even without input from the brain. However, the brain is the normal activator of these circuits. Therefore, when the spinal cord is transected, voluntary movements of the limbs cease. Also, without input from the brain, locomotor movements produced by spinal circuits are not capable of supporting the body or generating functional walking.

3. **True.** In the lamprey, each spinal segment has a central pattern generator on each side of the spinal cord. The circuit consists of motor neurons, two types of inhibitory interneurons, and one type of excitatory interneuron.

4. **True.** The axon of the crossed interneuron crosses the midline and makes inhibitory synapses onto the crossed interneuron, inhibitory interneuron, excitatory interneuron, and motor neurons of the contralateral central pattern generator.

5. **False.** Based on synaptic connections alone, alternation would not be expected. An important factor in terminating the burst of activity in the locomotor circuit is the decline in action potential rate in the excitatory interneuron, which results from accumulation of intracellular Ca^{2+}, not from a synaptic interaction.

6. **True.** The lamprey spinal cord contains stretch-sensitive sensory neurons that are stimulated by the bending of the spinal cord during swimming. The stretch-sensitive

neurons (called edge cells) have dendrites that spread laterally along the lateral margin of the cord. When muscles on the right side contract, the left side of the spinal cord becomes stretched and the edge cells on the left side fire action potentials. Both excitatory and inhibitory edge cells exist. Excitatory edge cells make excitatory synapses onto neurons of the ipsilateral central pattern generator. Inhibitory edge cells make inhibitory synapses onto neurons of the contralateral central pattern generator. When the right-side pattern generator is active, the body bends to the right and the left edge cells are activated. The inhibitory edge cell on the left then helps terminate activity in the right-side pattern generator, and the left-side excitatory edge cell helps initiate activity in the left-side pattern generator. Thus, the stretch-sensitive neurons in the spinal cord ensure correct alternation of motor output on the two sides of the body during swimming.

7. **False.** Interneurons in the spinal cord are used in various circuits, ranging from simple reflexes to complex motor tasks. The same interneurons used to extend one limb and flex the opposite limb during the withdrawal reflex might also participate in the same limb movements during locomotion.

8. **e.** Statement a is correct because the crossed interneuron is responsible for suppressing activity in neurons of the contralateral pattern generator circuit. Thus, as activity in the crossed interneuron declines, suppression declines and the contralateral circuit begins its burst. The mechanisms in statements b and c account for the decline of activity in the crossed interneuron.

IN THE CLINIC

Spinal reflexes are a mainstay of the basic neurological examination. Problems that affect the spinal cord can be detected by their effects on reflexes, often before more severe symptoms appear. Because reflexes involve both excitatory and inhibitory pathways in the spinal cord, hyperactive reflexes can also be a sign of spinal cord disease. The location of the affected muscles provides information about where in the spinal cord a lesion is located. In addition to reflex circuits, the spinal cord includes sensory pathways (Chapter 7) and descending motor pathways from the brain. Thus, spinal cord lesions often cause a complex mix of sensory and motor disturbances, which in combination can provide clues about not only the location of the lesion but also whether it affects primarily white matter or gray matter. Modern imaging techniques, such as magnetic resonance imaging, can then be used to confirm the location and extent of the pathological change.

Chapter Test

True/False

1. Like the myotatic reflex, the inverse myotatic reflex is activated by passive stretch of the muscle.

2. The myotatic reflex opposes passive stretch and thus maintains constant muscle length.

3. The withdrawal reflex requires interneurons whose axons cross the midline to the opposite side of the spinal cord, because the withdrawal reflex requires opposing actions on contralateral limbs.

4. Neural circuits for locomotion do not require synaptic interactions between the two sides of the spinal cord, because each side of the cord has independent central pattern generators.

5. Sensory neurons of the stretch reflex make contact with both extrafusal and intrafusal muscle fibers.

6. In the lamprey swimming circuit, timing of contraction on the two sides of the body is determined by the timing of the action potentials in motor command axons descending from the brain on each side.

7. In the lamprey swimming circuit, motor command axons descending from the brain on each side of the spinal cord make excitatory synapses only with motor neurons.

8. Gamma motor neurons control contraction of intrafusal muscle fibers, ensuring that sensitivity of the stretch-sensitive sensory neurons of the muscle spindle is maintained during active contraction.

9. In the withdrawal reflex, both inhibitory and excitatory interneurons are involved on both sides of the spinal cord.

10. The central pattern generator is a type of neuron in the brain that commands motor neurons of the spinal cord to fire at appropriate times during locomotion.

Multiple Choice

11. Which of the following reflexes involves a direct synapse between sensory neurons and motor neurons?
    a. inverse myotatic reflex
    b. withdrawal reflex
    c. myotatic reflex
    d. all of the above
    e. a and c but not b

12. During swimming in the lamprey,
    a. crossed interneurons on the two sides of the spinal cord fire alternating bursts of action potentials.
    b. the command axons on the two sides of the spinal cord fire alternating bursts of action potentials.
    c. the excitatory interneurons on the two sides of the spinal cord fire simultaneously and continuously.
    d. all of the above.
    e. none of the above.

13. During locomotion,
    a. sensory input from the limbs is required for alternating flexion/extension of each limb.
    b. sensory input from the limbs is required for normal locomotion.
    c. the withdrawal reflex is triggered in each limb in alternation.
    d. all of the above.
    e. none of the above.

14. A group of spinal inhibitory interneurons makes inhibitory synapses onto motor neurons controlling the flexor muscle of a joint. This group of inhibitory neurons could be part of
    a. the inverse myotatic reflex stimulated by sensory neurons of the tendon organs of the extensor muscle of the same joint.
    b. the myotatic reflex stimulated by sensory neurons of the muscle spindles from that same flexor muscle.
    c. the withdrawal reflex for the limb containing the flexor muscle.
    d. all of the above.
    e. none of the above.

15. The myotatic reflex and the inverse myotatic reflex are similar in which of the following ways?
    a. Both reflexes produce opposing effects on agonistic and antagonistic muscles.
    b. Both reflexes involve the activation of inhibitory interneurons in the spinal cord.
    c. Both reflexes involve motor neurons that feed back to the same muscle from which the sensory signal originated.
    d. All of the above.
    e. None of the above.

16. The circuitry of the withdrawal reflex includes
    a. nociceptor sensory neurons.
    b. excitatory interneurons on both sides of the spinal cord.
    c. inhibitory interneurons on both sides of the spinal cord.
    d. both flexor and extensor motor neurons.
    e. all of the above.

Short Answer

17. Match each of the following sensory receptors with the corresponding reflex:
    Receptors: muscle spindle, Golgi tendon organ, nociceptor.
    Reflexes: withdrawal, myotatic, inverse myotatic.

18. Identify the following: edge cell, crossed interneuron.

19. Identify the following: intrafusal muscle fiber, γ motor neuron.

Essay

20. Describe how the swimming circuitry in the lamprey spinal cord produces alternating bursts of action potentials in the motor neurons controlling the body muscles on the two sides of the body.

Chapter Test Answers

1. **F** 2. **T** 3. **T** 4. **F** 5. **F** 6. **F** 7. **F** 8. **T** 9. **T** 10. **F** 11. **c** 12. **a**
13. **b** 14. **e** 15. **d** 16. **e**

17. Muscle spindle: myotatic reflex; Golgi tendon organ: inverse myotatic reflex; nociceptor: withdrawal reflex.

18. Edge cell: stretch-sensitive sensory neuron in the spinal cord of the lamprey that modulates the activity of the central pattern generator for swimming. Excitatory edge cells excite neurons of the ipsilateral central pattern generator, and inhibitory edge cells inhibit the neurons of the contralateral central pattern generator. Crossed interneuron: a cell in the central pattern generator for swimming in the lamprey spinal cord. This type of interneuron inhibits activity of the contralateral central pattern generator during contraction of the ipsilateral body muscles.

19. Intrafusal muscle fiber: the muscle fibers contained within the muscle spindle. The stretch-sensitive sensory neurons make annulospiral endings around intrafusal muscle fibers. γ motor neurons: a special group of motor neurons that innervate the intrafusal muscle fibers and stimulate them to contract in parallel with the extrafusal muscle fibers.

20. A fish swims by alternating contractions of the body muscles, driven by alternating bursts of action potentials in the motor neurons on the two sides. The motor neurons are excited by synaptic inputs from the descending motor command fibers from the brain, which initiate swimming, and from excitatory interneurons of the ipsilateral central pattern generator. When the central pattern generator on the left side is active, crossed interneurons from the left side fire a burst of action potentials and inhibit neurons of the central pattern generator on the right side. This ensures that the right-side central pattern generator remains silent whereas the left-side central pattern generator is active, and the body thus bends to the left. The burst of action potentials in the left-side central pattern generator terminates because of three factors: excitation provided by the excitatory interneurons of the central pattern generator declines as intracellular Ca^{2+} builds up and activates K^+ channels, activity of inhibitory interneurons of the left-side central pattern generator reduces activity in the crossed interneuron from the left side, and inhibitory edge cells on the right side of the spinal cord are activated by the bending of the animal to the left, producing inhibition of the left-side central pattern generator. As activity in the left-side central pattern generator declines, neurons in the right-side central pattern generator are more likely to fire, because of declining inhibition from the crossed interneuron from the left side. Also, excitatory edge cells on the right side are activated by the bending of the animal to the left, and these excitatory edge cells add additional excitation to the neurons of the right-side central pattern generator. Once a burst of action potentials begins in the crossed interneuron from the right side, the activity in the left central pattern generator is rapidly terminated. Thus, muscles on the left side relax and muscles on the right side contract, bending the body toward the right. The sequence described above for the left-side central pattern generator then repeats for the central pattern generator on the right side.

Check Your Performance:

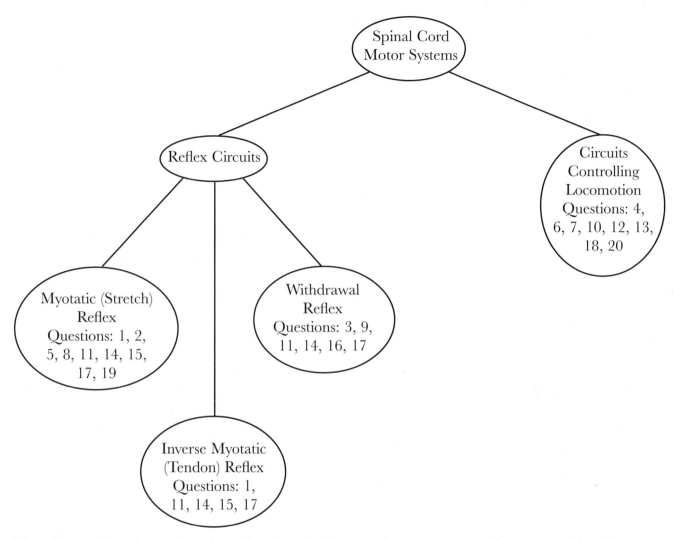

Note the number of questions in each grouping that you got wrong on the chapter test. Identify areas where you need further review and go back to relevant parts of this chapter.

Further help: If you continue to have difficulty with the concepts, a detailed presentation of these topics can be found in chapter 10 of *Neurobiology: Molecules, Cells, and Systems*, published by Blackwell Science, Inc.

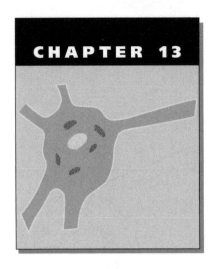

CHAPTER 13

Brain Motor Mechanisms

The brain stem contains a number of motor control circuits. In simple vertebrates, the brain stem is primarily responsible for guiding motor behavior. As the forebrain increased in size and complexity during evolution, motor circuits in the forebrain, especially the cerebral cortex, became increasingly important in motor control. The brain stem motor mechanisms were retained, forming an intermediate level between the cortex and the spinal cord (the **corticobulbar system**) in parallel with a direct route (the **corticospinal system**) from the cortex directly to the spinal cord. Although exceptions exist, brain stem motor systems tend to govern global motions, such as locomotion, and the corticospinal system provides finer motor control, such as control of the fingers in primates.

ESSENTIAL BACKGROUND

- **Anatomical organization of the forebrain**
- **Anatomical organization of the brain stem and cerebellum**
- **Spinal cord motor systems**

TOPIC 1: CORTICAL AND BRAIN STEM MOTOR SYSTEMS

KEY POINTS

✓ *What parts of the cerebral cortex give rise to the corticospinal and corticobulbar tracts?*

✓ *What is the somatotopic organization of the primary motor cortex?*

✓ *What descending tracts from the brain stem are involved in control of spinal motor circuits?*

Figure 13.1A summarizes the major components of the brain motor system. The motor cortex includes the primary motor cortex, supplementary motor cortex, and the premotor cortex. The **primary motor cortex** is located in the **precentral gyrus**. Descending axons of pyramidal cells in layer V of the primary motor cortex form the **corticospinal tract** (see Figure 13.1A), which provides direct control of spinal motor circuits. In the medulla, corticospinal axons form large bundles, called the **pyramids**. The primary motor cortex has a somatotopic organization,

with the most lateral parts controlling the head and mouth, the most medial parts controlling the hindlimb, and the intermediate parts controlling the trunk and forelimb. The right motor cortex controls the left side of the body, and the left motor cortex controls the right side. Anterior to

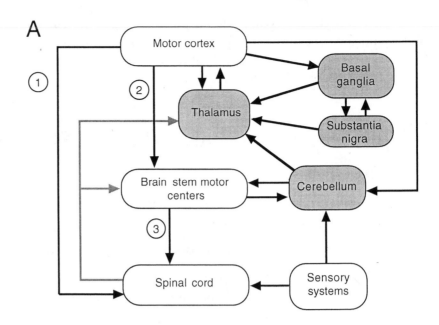

A

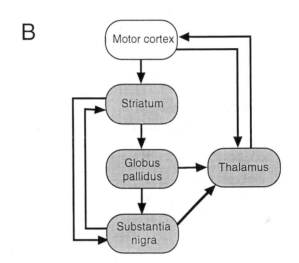

B

Figure 13.1 The organization of motor control systems in the central nervous system. (A) The overall scheme. Gray arrows indicate sensory information. The arrow labeled 1 represents the corticospinal tract, the arrow labeled 2 represents the corticobulbar tract, and the arrow labeled 3 represents the combined reticulospinal, vestibulospinal, and rubrospinal tracts. (B) The detailed connections among the cerebral cortex, basal ganglia, and motor nuclei of the thalamus.

primary motor cortex are two additional cortical motor areas, the **supplemental motor area** and the **premotor cortex**. Neurons in these areas send axons to primary motor cortex and to brain stem and spinal motor circuits. The supplemental motor area and the premotor cortex program global movements that involve coordinated action of different muscles of an appendage (i.e., moving all fingers on one hand). By contrast, neurons of primary motor cortex control simpler movements (i.e., moving a single finger).

The motor cortical areas also send axons to the brain stem in the **corticobulbar tract** (see Figure 13.1A). The brain stem gives rise to three major groups of descending axons (see Figure 13.1A) in parallel with the corticospinal tract. First, the **reticulospinal tract** originates in the reticular formation and makes synaptic connections onto motor neurons and interneurons throughout the spinal cord. Excitatory inputs from reticulospinal neurons in the brain stem drive the central pattern generators of the spinal cord during locomotion (Chapter 12). Second, the **vestibulospinal tract** originates in the vestibular nuclei of the brain stem and promotes extension of the limbs and inhibits flexion. Third, the **rubrospinal tract** originates in the red nucleus and promotes flexion of the limbs and inhibits extension. The balance of activity in the vestibulospinal system and rubrospinal system is important in governing the flexion/extension equilibrium of the limbs.

Topic Test 1: Cortical and Brain Stem Motor Systems

True/False

1. Some neurons in the primary motor cortex have axons that extend all the way to the spinal cord, without intervening synapses.

2. The primary motor cortex is in the postcentral gyrus, just behind the central sulcus.

3. The primary motor cortex has a somatotopic organization, with the lower limb represented near the midline and the head at the lateral edge.

4. A lesion in the left motor cortex will interfere with movements of the right side of the body.

5. The primary motor cortex is the only part of the cortex that sends outputs directly to the spinal and brain stem motor circuits.

6. Although brain stem motor nuclei are sufficient to coordinate locomotion in simple vertebrates, cortical motor areas are required for coordinated locomotion in animals with more complex brains.

7. The reticulospinal tract is part of the corticospinal tract.

8. The vestibulospinal tract and reticulospinal tract have opposing effects on flexion and extension of the limbs.

Multiple Choice

9. Which of the following is part of the descending motor system from the brain stem to the spinal cord?
   a. the pyramidal tract
   b. the corticobulbar tract
   c. the rubrospinal tract
   d. the corticospinal tract
   e. none of the above

10. The corticospinal tract
    a. provides finer motor control than the more global effects produced by brain stem motor systems.
    b. bypasses brain stem motor systems and makes synapses directly onto spinal neurons.
    c. originates from pyramidal neurons found in layer V of the motor cortex.
    d. all of the above.
    e. none of the above.

Topic Test 1: Answers

1. **True.** Axons of output neurons of primary motor cortex (pyramidal cells in layer V) form the corticospinal tract, which descends from the cortex directly into the spinal cord and makes synapses onto spinal neurons.

2. **False.** The primary motor cortex is in the precentral gyrus, in front of the central sulcus.

3. **True.** Note that the primary motor cortex and primary somatosensory cortex (Chapter 7) have the same somatotopic representation of the body, arranged from medial to lateral along the cortex. Thus, sensory inputs from a particular part of the body are adjacent to the motor command neurons for that same part of the body.

4. **True.** Descending axons of the corticospinal tract decussate as they project through the medulla to the spinal cord. The crossover point in the medulla is called the pyramidal decussation. Thus, cortical neurons on each side of the brain connect with spinal motor neurons controlling the opposite side of the body.

5. **False.** The supplemental motor area and the premotor cortex also have outputs to the brain stem and spinal cord. In addition, these cortical areas send axons to the primary motor cortex. Therefore, the supplemental motor area and premotor cortex have both direct and indirect influence on brain stem and spinal motor circuits.

6. **False.** Even when the cortex is completely removed from the mammalian brain, locomotor movements can still be produced. Electrical stimulation of the midbrain induces walking, trotting, or galloping on a treadmill in a decerebrate animal, depending on the strength of electrical stimulation. This trigger region for locomotion is at the upper end of the reticular formation.

7. **False.** The reticulospinal tract projects from the reticular formation in the brain stem to the spinal cord. In simple vertebrates (e.g., the lamprey), this tract constitutes the principal control pathway from the brain to the spinal cord. The reticulospinal projection still exists in animals with well-developed forebrains, but it is supplemented by cortico-spinal projections.

8. **False.** The reticulospinal tract is involved in initiation of locomotion, which involves alternating flexion and extension of the limbs. The vestibulospinal tract promotes limb extension, and the rubrospinal tract promotes limb flexion. Thus, the vestibulospinal tract and rubrospinal tract have opposite effects on flexion and extension.

9. **c.** The rubrospinal tract extends from the red nucleus of the brain stem to the spinal cord. Neurons of the red nucleus make excitatory synapses onto spinal interneurons that in turn excite flexor motor neurons of the limbs. The pyramidal tract is another name for the corticospinal tract, containing descending axons of cortical neurons to the spinal cord. The corticobulbar tract carries axons from cortical neurons to the brain stem.

10. **d.** Brain stem motor systems typically control global motor actions, such as alternating swinging of limbs during walking in tetrapods. The corticospinal tract typically produces fine movements, such as delicate manipulation of objects with the fingers. The corticospinal tract contains the axons of pyramidal neurons of layer V of the motor cortex, which pass through the brain stem and terminate directly on spinal neurons.

TOPIC 2: BASAL GANGLIA, THALAMUS, AND CEREBELLUM

KEY POINTS

✓ *What are the basal ganglia?*

✓ *What are the interconnections between the basal ganglia and the motor cortex?*

✓ *How does the cerebellum influence motor outputs from the brain to the spinal cord?*

The **basal ganglia** in the forebrain receive inputs from the motor cortex and project back to the motor cortex indirectly via thalamic nuclei. The basal ganglia are divided into the **globus pallidus** and **striatum**, and the striatum is further subdivided into the **caudate nucleus** and **putamen**. The basal ganglia are also closely associated with a midbrain region called the **substantia nigra**. Together with the thalamus, the basal ganglia and substantia nigra form three interconnected feedback loops, shown in Figure 13.1A. First, a local feedback loop runs from the basal ganglia to the substantia nigra and then back to the basal ganglia. Second, a loop extends from motor cortex, to the basal ganglia, to the thalamus, and then back to motor cortex. Another slightly longer path leads from motor cortex, to the basal ganglia, to the substantia nigra, to the thalamus, and then back to motor cortex. These interrelated feedback loops allow the basal ganglia and the substantia nigra, acting via the thalamus, to influence cortical motor outputs. A more detailed view of the connections among the cortex, basal ganglia, substantia nigra, and thalamus is given in **Figure 13.1B**.

Another brain motor area is the **cerebellum** of the brain stem. The cerebellum consists of two principal divisions, the **deep cerebellar nuclei** and the **cerebellar cortex**, both of which receive a variety of sensory information and copies of motor commands sent by cortical and brain stem motor centers. The deep nuclei send feedback directly to brain stem motor centers and indirectly (via the thalamus) to cortical centers. The cerebellum may have evolved as a brain stem center where vestibulospinal and reticulospinal motor outputs are combined with vestibular information to adjust motor commands during swimming. Fine coordination of limb movements and maintenance of balance during locomotion are among the functions of the cerebellum in mammals. The cerebellum also plays a role in motor learning.

Topic Test 2: Basal Ganglia, Thalamus, and Cerebellum

True/False

1. Neurons of the basal ganglia send axons directly to motor cortex, where they synapse onto output neurons of the corticospinal tract.

2. The three nuclei making up the basal ganglia are the striatum, globus pallidus, and substantia nigra.

3. The substantia nigra receives inputs from the striatum and sends a reciprocal connection back to the striatum.

4. The basal ganglia receive a direct synaptic input from primary motor cortex.

5. The motor nuclei of the thalamus send a direct synaptic connection to motor cortex, but the thalamus does not receive synapses from neurons of the motor cortex.

6. The cerebellum consists two convoluted hemispheres overlying the top of the brain stem and three deep nuclei: the dentate nucleus, emboliform nucleus, and fastigial nucleus.

7. Axons of the corticothalamic projection and the corticospinal projection originate from the same neurons in the motor cortex.

Multiple Choice

8. The motor nuclei of the thalamus receive direct synaptic connections from
   a. the cerebellum.
   b. the motor cortex.
   c. the globus pallidus.
   d. a and b but not c.
   e. a, b, and c.

Topic Test 2: Answers

1. **False.** The basal ganglia do not project directly to the motor cortex. Instead, they send axons to thalamic nuclei that in turn provide inputs to the motor cortex. Thus, the basal ganglia influence cortical motor output indirectly by modulating thalamic inputs to the cortex.

2. **False.** The three nuclei of the basal ganglia are the globus pallidus, caudate nucleus, and putamen. Together, the caudate nucleus and the putamen make up the striatum. Unlike the basal ganglia, which are part of the telencephalon, the substantia nigra is located in the midbrain.

3. **True.** Neurons of the substantia nigra receive inputs from and send outputs to the striatum. In addition, different substantia nigra neurons send an inhibitory connection to the thalamic nuclei that relay outputs from the basal ganglia to the motor cortex.

4. **True.** Neurons in the motor cortex send axons to the basal ganglia, where they make synapses on neurons of the striatum.

5. **False.** Motor nuclei of the thalamus send outputs to and receive inputs from motor cortex. Thus, the thalamus is an important part of feedback loops involving the motor cortex, basal ganglia, and substantia nigra.

6. **True.** In animals with a highly developed cerebellum, the cerebellum is a large, highly convoluted pair of hemispheres tucked between the posterior pole of the cerebrum and the core of the brain stem. The cerebellar hemispheres have an organization analogous to that of the cerebral hemispheres, with a densely packed cellular layer (the cerebellar cortex) on the outside and white matter consisting of incoming and outgoing axons underneath. The deep nuclei (the dentate nucleus, emboliform nucleus, and fastigial nucleus) are hidden from view between the overlying cerebellar hemispheres and the rest of the brain stem. The cerebellar hemispheres and deep nuclei receive copies of motor commands originating from cortical and brain stem motor circuits and also receives various sensory signals. The cerebellum modulates motor commands issued by other brain regions, based on sensory feedback about the outcome of motor performance.

7. **False.** The corticospinal tract originates from neurons in layer V of motor cortex. Layer VI of the cortex contains neurons that send axons to motor parts of the thalamus. Thus, the corticospinal and corticothalamic projections originate from different populations of neurons.

8. **e.** All motor areas described in a through c make direct synaptic connections with the motor nuclei of the thalamus. The globus pallidus is the part of the basal ganglia that projects to the thalamus.

IN THE CLINIC

The importance of the substantia nigra and the basal ganglia in motor behavior is clearly demonstrated by Parkinson's disease, a condition associated with muscle tremors and difficulty in initiating and sustaining locomotion. Patients sometimes "freeze up" during locomotion and report that motor intentions do not get translated into motor performance. Parkinson's disease is caused by degeneration of neurons in the substantia nigra that project to the striatum (the nigrostriatal projection). Loss of this synaptic input severely impairs motor activity by disrupting the normal feedback loops involving the basal ganglia. Nigrostriatal neurons release the neurotransmitter dopamine, and drugs that mimic the missing dopamine or potentiate the action of remaining dopamine reduce the symptoms of Parkinson's disease. The most commonly used treatment is L-dopa, which is the immediate precursor of dopamine in the synthesis pathway for the neurotransmitter.

Chapter Test

True/False

1. The supplemental motor area of the cortex projects to the brain stem and spinal cord only indirectly, by means of connections to primary motor cortex.

2. The basal ganglia affect motor output from the cortex indirectly, by means of synaptic interactions with thalamic motor nuclei.

3. The reticulospinal tract is involved in triggering motor circuits of locomotion in the spinal cord.

4. The rubrospinal system promotes extension of the limbs, and the vestibulospinal system promotes flexion of the limbs.

5. The primary motor cortex is located in the precentral gyrus of the cerebral cortex.

6. The corticobulbar tract bypasses brain stem motor systems and projects directly from the cortex to the spinal cord.

7. The deep cerebellar nuclei receive inputs from cortical motor areas but not from brain stem motor centers.

8. The dentate nucleus is one of the deep nuclei of the cerebellum.

9. Neurons of the substantia nigra receive synaptic inputs from the motor nuclei of the thalamus and the motor cortex.

10. The medial part of the primary motor cortex on the right side of the brain controls muscles of the lower limb on the left side of the body.

Multiple Choice

11. Which of the following motor acts would be most dependent on the corticospinal motor system?
    a. playing the piano
    b. walking
    c. swinging a baseball bat
    d. balancing on one leg
    e. removing the hand from a hot object

12. The cerebellum receives synaptic input from
    a. the motor nuclei of the thalamus.
    b. brain stem motor centers.
    c. the substantia nigra.
    d. all of the above.
    e. none of the above.

13. The three deep nuclei of the cerebellum are
    a. the striatum, red nucleus, and vestibular nucleus.
    b. the caudate nucleus, substantia nigra, and pyramidal nucleus.
    c. the precentral gyrus, putamen, and red nucleus.
    d. the dentate nucleus, emboliform nucleus, and fastigial nucleus.
    e. none of the above.

14. Which of the following regions gives rise to a descending motor tract to the spinal cord?
    a. cerebellum
    b. substantia nigra
    c. red nucleus
    d. all of the above
    e. none of the above

15. The motor nuclei of the thalamus
    a. provide the only feedback route to the motor cortex from noncortical motor centers.
    b. send axons to all cortical and subcortical motor centers of the brain, including brain stem motor systems.
    c. receive ascending axons from the red nucleus and vestibular nucleus.
    d. all of the above.
    e. none of the above.

16. Neurons of the motor cortex send axons to
    a. the spinal cord.
    b. brain stem motor centers.
    c. the striatum.
    d. all of the above.
    e. none of the above.

Short Answer

17. List the three descending tracts carrying motor signals from the brain stem to the spinal cord.

18. List the three nuclei that make up the basal ganglia.

19. List four motor centers that send direct synaptic connections to the thalamus.

Essay

20. Describe two feedback loops involving the motor cortex and the basal ganglia.

Chapter Test Answers

1. **F**   2. **T**   3. **T**   4. **F**   5. **T**   6. **F**   7. **F**   8. **T**   9. **F**   10. **T**   11. **a**   12. **b**

13. **d**   14. **c**   15. **a**   16. **d**

17. Reticulospinal tract, rubrospinal tract, and vestibulospinal tract.

18. Globus pallidus, caudate nucleus, and putamen.

19. Motor cortex, basal ganglia (globus pallidus), substantia nigra, and deep cerebellar nuclei.

20. One feedback loop originates in the motor cortex, which projects to the basal ganglia. Neurons in the basal ganglia then send outputs to the motor nuclei of the thalamus, which then send axons back to the motor cortex. Within this feedback loop, the striatum receives inputs from the motor cortex in the basal ganglia and then projects to the globus pallidus. The globus pallidus then sends axons to the thalamus. A second feedback loop also begins with the projection from motor cortex to the striatum, but the neurons of the striatum then send axons to the substantia nigra. A subgroup of neurons in the substantia nigra project to the thalamus, which in turn completes the loop back to the motor cortex. Other neurons in the substantia nigra (the dopamine-releasing neurons) send axons back to the striatum, completing a local feedback loop between the basal ganglia and the substantia nigra.

Check Your Performance:

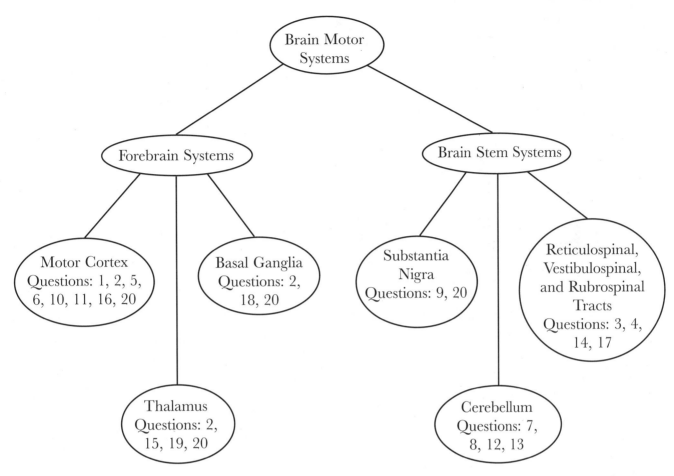

Note the number of questions in each grouping that you got wrong on the chapter test. Identify areas where you need further review and go back to relevant parts of this chapter.

Further help: If you continue to have difficulty with the concepts, a detailed presentation of these topics can be found in chapter 11 of *Neurobiology: Molecules, Cells, and Systems*, published by Blackwell Science, Inc.

CHAPTER 14

The Autonomic Nervous System

The **somatic nervous system** controls skeletal muscles and the **autonomic nervous system** controls the cardiovascular system, respiratory system, digestive system, and other organ systems involved in internal homeostasis.

ESSENTIAL BACKGROUND

- Anatomy of the vertebrate body
- Physiology of the mammalian heart
- Anatomy of the brain and spinal cord

TOPIC 1: SYMPATHETIC AND PARASYMPATHETIC DIVISIONS OF THE AUTONOMIC NERVOUS SYSTEM

KEY POINTS

✓ *What neurotransmitters are used by the motor neurons of the sympathetic and parasympathetic divisions of the autonomic nervous system?*

✓ *How do the neurotransmitters of the autonomic nervous system affect target organs?*

✓ *What parts of the central nervous system control autonomic ganglia?*

General organization of the autonomic nervous system is described in Chapter 1. Parasympathetic motor neurons release acetylcholine, which typically acts on muscarinic cholinergic receptors to alter the level of second messengers (e.g., cyclic AMP) in target cells. Sympathetic motor neurons release the transmitter norepinephrine (also called noradrenaline), which also alters second messenger levels. In general, sympathetic activity makes the organism ready for vigorous activity ("fight or flight"), and the parasympathetic nervous system favors reduced activity ("rest and digest"). The central nervous system controls autonomic ganglia via output neurons called **preganglionic neurons**. Parasympathetic preganglionic neurons are located in the brain stem and sacral spinal cord. Parasympathetic preganglionic fibers exit the spinal cord in the **splanchnic nerve**. In the brain, parasympathetic preganglionic fibers exit via cranial nerves III (**oculomotor nerve**), VII (**facial nerve**), IX (**glossopharyngeal nerve**), and X (**vagus nerve**). Sympathetic preganglionic neurons are located in the **intermediolateral gray matter** of the spinal cord, from thoracic to lumbar levels. Both sympathetic and parasympathetic preganglionic neurons release acetylcholine.

Topic Test 1: Sympathetic and Parasympathetic Divisions of the Autonomic Nervous System

True/False

1. Motor neurons of the autonomic nervous system do not control muscle cells.

2. Motor neurons (postganglionic neurons) release the neurotransmitter norepinephrine in the sympathetic division and acetylcholine in the parasympathetic division.

3. All sympathetic preganglionic neurons are located in the spinal cord.

4. All parasympathetic preganglionic neurons are located in the spinal cord.

Multiple Choice

5. Which of the following neurons release acetylcholine?
   a. somatic motor neurons
   b. parasympathetic motor neurons
   c. parasympathetic preganglionic neurons
   d. sympathetic preganglionic neurons
   e. all of the above

6. Neurons of the intermediolateral cell column make synaptic contact with which of the following?
   a. paravertebral sympathetic ganglia
   b. prevertebral sympathetic ganglia
   c. sympathetic chain ganglia
   d. the adrenal gland
   e. all of the above

7. In the sympathetic nervous system, norepinephrine usually acts on postsynaptic target cells
   a. by directly opening ion channels in the postsynaptic membrane.
   b. by activating G protein-coupled receptors, which indirectly affect ion channels by altering second messenger levels.
   c. by directly binding to voltage-dependent ion channels in the postsynaptic cell, preventing the channels from opening upon depolarization.
   d. a and b but not c.
   e. a, b, and c.

Topic Test 1: Answers

1. **False.** Autonomic motor neurons do not contact skeletal muscle cells but do contact other kinds of muscle cells, such as smooth muscle cells of blood vessels. Striated muscle cells of the heart are also regulated by the autonomic nervous system.

2. **True.** Sympathetic and parasympathetic postganglionic (motor) neurons have opposing actions on target cells, mediated by two different neurotransmitters. Sympathetic postganglionic cells release norepinephrine, and parasympathetic postganglionic neurons release acetylcholine.

3. **True.** Sympathetic preganglionic neurons are located in the intermediolateral gray matter in the thoracic and upper lumbar segments of the spinal cord.

4. **False.** Some parasympathetic preganglionic neurons are located in the sacral spinal cord, but most parasympathetic preganglionic neurons are located in cranial nerve nuclei of the midbrain, pons, and medulla.

5. **e.** Acetylcholine is released by preganglionic neurons of both divisions of the autonomic nervous system. Somatic motor neurons also release acetylcholine, as do parasympathetic postganglionic neurons.

6. **e.** The intermediolateral cell column contains sympathetic preganglionic neurons. Most axons of these preganglionic neurons exit the spinal cord and enter the paravertebral chain of sympathetic ganglia that runs parallel to the spinal column, making synapses onto postganglionic neurons in the paravertebral sympathetic ganglia. The paravertebral ganglia are also called the sympathetic chain ganglia, or simply the sympathetic chains. Some sympathetic preganglionic neurons send axons to the prevertebral sympathetic ganglia, which innervate the abdomen and sex organs. Another target of sympathetic preganglionic neurons is the adrenal gland, which contains endocrine cells called chromaffin cells. Chromaffin cells are modified sympathetic postganglionic cells, which release norepinephrine and epinephrine into the bloodstream.

7. **b.** The postsynaptic action of norepinephrine at sympathetic targets is usually mediated by receptors coupled indirectly to ion channels. An example is the effect of norepinephrine on the heart, where sympathetic input speeds heart rate and increases strength of contraction. This effect is mediated by increased opening of voltage-dependent Ca^{2+} channels during depolarization. The norepinephrine receptor molecules of the heart are beta-adrenergic receptors, which activate G proteins. G proteins in turn stimulate the synthetic enzyme for cyclic AMP, and the level of cyclic AMP increases. Cyclic AMP stimulates protein kinase A, which phosphorylates voltage-dependent Ca^{2+} channels and potentiates their opening upon depolarization.

TOPIC 2: BRAIN SYSTEMS CONTROLLING THE AUTONOMIC NERVOUS SYSTEM

KEY POINTS

✓ *What parts of the brain influence autonomic preganglionic neurons?*

✓ *What synaptic circuitry underlies the cardiac baroreceptor reflex?*

As in the somatic nervous system, reflex loops are important in autonomic motor systems. However, in the autonomic nervous system, sensory inputs reach the motor outputs only after several synaptic relay stages, and the reflex loop often goes through the brain stem. An example is the cardiac **baroreceptor reflex** (**Figure 14.1**). When blood pressure rises, stretch receptor neurons are activated in the carotid sinus (a specialized part of the carotid artery) and in the aorta (the main artery leaving the heart). These stretch receptors reflexively lower blood pressure by simultaneously stimulating parasympathetic input and inhibiting sympathetic input to the heart. Axons of the arterial stretch receptor neurons enter the brain in the medulla and make excitatory synapses on neurons in the nucleus of the solitary tract. This nucleus is a multipurpose area that receives inputs from various sensory systems and regulates homeostatic functions. After passing through one or more excitatory interneurons in the **nucleus of the solitary tract**, excitation is relayed to parasympathetic preganglionic neurons of the heart, in

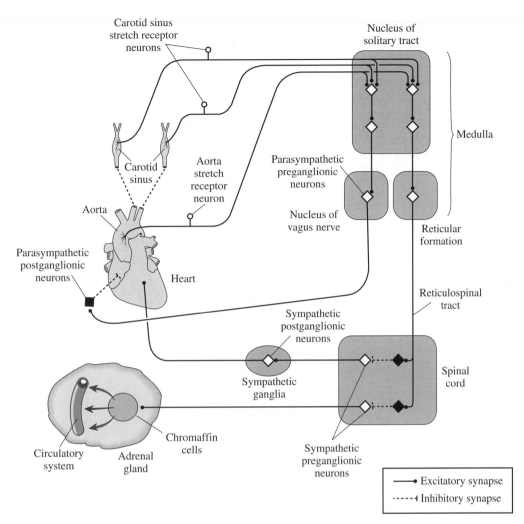

Figure 14.1 The cardiac baroreceptor reflex. Stretch receptor neurons innervate the arterial walls of the aorta and of specialized regions of the carotid arteries, the carotid sinuses. These stretch receptors are activated by stretch of the artery wall produced by the blood pressure. The stretch receptor neurons make excitatory connections onto neurons within the nucleus of the solitary tract in the medulla. These neurons then project via additional interneurons to the parasympathetic preganglionic neurons of the heart, located in the nucleus of the vagus nerve, and to reticulospinal neurons that send axons to the spinal cord. Increased activity of the stretch receptors excites the parasympathetic preganglionic neurons, which in turn excite the parasympathetic neurons of the heart and cause increased release of acetylcholine. The reticulospinal neurons excite interneurons in the spinal cord that inhibit sympathetic preganglionic neurons involved in control of the heart.

the nucleus of the vagus nerve. The preganglionic neurons then excite cardiac parasympathetic postganglionic neurons, which release acetylcholine and inhibit cardiac activity.

Other interneurons in the nucleus of the solitary tract make excitatory synapses onto reticulospinal neurons, which excite spinal interneurons that inhibit sympathetic preganglionic neurons. This reduces sympathetic input to the heart, decreasing strength of contraction and blood pressure. In addition, preganglionic neurons that stimulate adrenal chromaffin cells are also inhibited in the spinal cord. The release of epinephrine and norepinephrine into the blood from the adrenal gland is thus also reduced, which further decreases cardiac output.

Many functions carried out by the autonomic nervous system are controlled by the **hypothalamus**, a collection of nuclei surrounding the third ventricle at the base of the diencephalon.

Because of its importance in controlling the autonomic nervous system, the hypothalamus is sometimes referred to as the head ganglion of the autonomic nervous system. Stimulation of the lateral hypothalamus, for example, produces general activation of the sympathetic nervous system, similar to that occurring in a "flight-or-fight" situation.

Topic Test 2: Brain Systems Controlling the Autonomic Nervous System

True/False

1. Autonomic reflexes can involve a direct synaptic connection between primary sensory neurons and preganglionic neurons.

2. The cardiac baroreceptor reflex is initiated by activation of stretch-sensitive sensory neurons.

3. In the baroreceptor reflex, preganglionic neurons in the nucleus of the vagus nerve are inhibited when blood pressure increases.

4. In the baroreceptor reflex, preganglionic neurons in the intermediolateral cell column are inhibited when blood pressure increases.

5. The reticulospinal pathway is involved in autonomic reflexes.

6. The cardiac baroreceptor reflex can either increase or decrease cardiac output.

7. The axons of arterial stretch receptor neurons of the baroreceptor reflex enter the brain in the medulla, where they make excitatory synapses onto interneurons of the hypothalamus.

Multiple Choice

8. The hypothalamus
   a. is a collection of nuclei located in the basal part of the diencephalon.
   b. includes neurons that induce an overall activation of the sympathetic nervous system.
   c. is part of the thalamus.
   d. a and b but not c.
   e. a, b, and c.

Topic Test 2: Answers

1. **False.** Autonomic reflexes typically involve many synaptic stages between sensory neurons and preganglionic neurons.

2. **True.** When blood pressure rises, the walls of blood vessels expand, activating stretch-sensitive sensory neurons of the carotid sinus and aorta. These sensory neurons initiate the baroreceptor reflex and reduce cardiac output.

3. **False.** Preganglionic neurons in the nucleus of the vagus nerve include the parasympathetic preganglionic neurons of the heart. In the baroreceptor reflex, the cardiac preganglionic neurons are excited by stretch of the aorta and carotid sinus, slowing the heart rate and decreasing cardiac output.

4. **True.** The intermediolateral cell column of the spinal cord contains sympathetic preganglionic neurons, including those that control sympathetic innervation of the heart. When blood pressure increases and the stretch-sensitive sensory neurons of the aorta and carotid sinus are activated, excitation is passed along from the medullary reticular formation to inhibitory interneurons in the spinal cord. These interneurons in turn inhibit the sympathetic preganglionic neurons of the heart.

5. **True.** Reticulospinal neurons form the pathway from the medulla to the spinal cord.

6. **True.** Even at normal blood pressure, the arteries are somewhat stretched, so a certain level of activity usually occurs in the stretch receptor neurons. This ongoing activity contributes steady excitation to the cardiac parasympathetic pathway and steady inhibition to the cardiac sympathetic pathway, via the synaptic connections of the baroreceptor reflex. In addition, other sources contribute excitatory and inhibitory inputs to both the sympathetic and parasympathetic pathways. The result is that both sympathetic and parasympathetic inputs to the heart are active at normal blood pressure, and heart rate and cardiac output at any moment represent the sum of these competing influences. When arterial pressure falls below the normal level, decreased activity of the stretch receptors reduces both excitation reaching the parasympathetic preganglionic neurons and inhibition reaching the sympathetic preganglionic neurons.

7. **False.** The synaptic target of the arterial stretch receptor neurons of the baroreceptor reflex is the nucleus of the solitary tract in the medulla. The hypothalamus is part of the diencephalon, in the forebrain.

8. **d.** The hypothalamus is not part of the thalamus. Neurons in the lateral hypothalamus do indeed produce overall activation of the sympathetic nervous system.

IN THE CLINIC

Many clinically useful drugs target the cholinergic and adrenergic receptors used by parasympathetic and sympathetic postganglionic neurons to regulate organ systems. A premier example is the cardiovascular system. In patients with high blood pressure (hypertension), treatment commonly includes drugs (e.g., propranolol) that block beta-adrenergic receptors of the heart, reducing cardiac output. One common cause of hypertension is stress, which causes an overall increase in activity of the sympathetic division of the autonomic nervous system. An abnormally high set point of the baro-receptor reflex might underlie some cases of high blood pressure, causing the reflex circuitry to regulate blood pressure at too high a level. This abnormal set point might arise from synaptic influences in the central nervous system or from heightened sensitivity of the arterial stretch receptor neurons to stretch of the vessel walls. Another example of drugs that target autonomic postsynaptic receptors is the blockade of muscarinic acetylcholine receptors to produce dilation of the pupils (mydriasis), which facilitates ophthalmological examination of the retina. The diameter of the pupil is controlled by two antagonistic sets of muscles: circularly arranged sphincter muscles that constrict the pupil and radial dilator muscles that increase the pupillary opening. The constrictor muscles are activated by the parasympathetic input to the iris, and the dilator muscles are activated by the sympathetic input. Blocking the muscarinic receptors of the constrictor muscles therefore dilates the pupil.

Chapter Test

True/False

1. All autonomic preganglionic neurons release the neurotransmitter acetylcholine.

2. All autonomic preganglionic neurons are located in the central nervous system.

3. All parasympathetic preganglionic neurons are located in the brain.

4. Motor neurons of the autonomic nervous system are also called postganglionic neurons.

5. Motor neurons of the autonomic nervous system give rise to two nerve fibers, one of which projects to the central nervous system where it receives synapses from the preganglionic neurons.

6. In the autonomic nervous system, norepinephrine and acetylcholine usually have opposite actions on target cells because norepinephrine inhibits acetylcholine receptor molecules of the postsynaptic cell.

7. The nucleus of the solitary tract is a multipurpose nucleus in the medulla involved in regulation of the cardiovascular system.

8. The nucleus of the vagus nerve receives direct synaptic inputs from stretch-sensitive sensory neurons of the carotid sinus.

9. The hypothalamus is a collection of nuclei located near the third ventricle at the base of the diencephalon.

10. The hypothalamus coordinates cardiovascular reflexes but has no effect on other functions of the autonomic nervous system.

Multiple Choice

11. Autonomic and somatic motor neurons differ in the following way:
    a. all somatic motor neurons release the neurotransmitter acetylcholine, but all autonomic motor neurons release the neurotransmitter norepinephrine.
    b. all somatic motor neurons contact skeletal muscle cells, but all autonomic motor neurons contact smooth muscle cells.
    c. all somatic motor neurons have cell bodies in the gray matter of the spinal cord, but all autonomic motor neurons have cell bodies in ganglia located outside the central nervous system.
    d. all of the above.
    e. none of the above.

12. Which of the following neurons releases the neurotransmitter norepinephrine?
    a. parasympathetic postganglionic neuron
    b. parasympathetic preganglionic neuron
    c. sympathetic preganglionic neuron
    d. sympathetic postganglionic neuron
    e. none of the above

13. In the target tissues of the autonomic nervous system, acetylcholine typically acts by
    a. directly opening nonspecific cation channels in the postsynaptic cell.
    b. activating postsynaptic receptor molecules that are coupled to G proteins.

c. activating beta-adrenergic receptors.

d. all of the above.

e. none of the above.

14. In the baroreceptor reflex, a fall in blood pressure below its normal level would
    a. have no effect on heart rate or cardiac output.
    b. reduce activity in the parasympathetic innervation of the heart.
    c. increase activity in the sympathetic innervation of the heart.
    d. a and b but not c
    e. b and c but not a

15. The cardiac baroreceptor reflex involves
    a. the nucleus of the solitary tract.
    b. the nucleus of the vagus nerve.
    c. reticulospinal projections.
    d. all of the above.
    e. none of the above.

16. The hypothalamus is called the "head ganglion" of the autonomic nervous system because
    a. all sensory information from the body passes through the hypothalamus before distribution to the other parts of the autonomic nervous system.
    b. all autonomic reflexes involve neurons of the hypothalamus.
    c. the hypothalamus coordinates the actions of other autonomic centers that control individual organs and mediate specific reflexes.
    d. all of the above.
    e. none of the above.

Short Answer

17. Identify carotid sinus and nucleus of the solitary tract.

18. List the cranial nerve nuclei that contain parasympathetic preganglionic neurons.

19. What parts of the spinal cord contain autonomic preganglionic neurons?

Essay

20. Describe the cellular actions of the sympathetic neurotransmitter in cardiac muscle cells.

Chapter Test Answers

1. **T** 2. **T** 3. **F** 4. **T** 5. **F** 6. **F** 7. **T** 8. **F** 9. **T** 10. **F** 11. **c** 12. **d**

13. **b** 14. **e** 15. **d** 16. **c**

17. Carotid sinus: a region of the carotid artery specialized for sensing blood pressure by means of stretch-sensitive sensory neurons that detect expansion of the arterial wall as pressure rises. Nucleus of the solitary tract: a nucleus in the medulla that receives excitatory inputs from sensory neurons that detect stretch of the arterial walls. Neurons in the nucleus then make excitatory synapses onto cardiac parasympathetic preganglionic neurons in the nucleus of the vagus nerve and onto reticulospinal neurons.

18. Oculomotor nerve (cranial nerve III), facial nerve (cranial nerve VII), glossopharyngeal nerve (cranial nerve IX), and vagus nerve (cranial nerve X).

19. Sympathetic preganglionic neurons are found in the thoracic and upper lumbar segments in the intermediate part of the spinal cord gray matter (the intermediolateral cell column). Parasympathetic preganglionic neurons in the spinal cord are located in the sacral spinal cord.

20. Norepinephrine released from the sympathetic postganglionic neurons binds to beta-adrenergic receptors of cardiac muscle cells. When norepinephrine binds, the receptor activates G proteins by catalyzing the replacement of GDP by GTP on the alpha subunit of the G protein. In its GTP-bound form, the alpha subunit is then able to bind to and activate the enzyme adenylyl cyclase, which synthesizes the internal messenger cyclic AMP from ATP. Thus, norepinephrine increases the intracellular concentration of cyclic AMP in cardiac muscle cells. Cyclic AMP stimulates another enzyme called protein kinase A (also known as cAMP-dependent protein kinase), which in turn uses ATP to phosphorylate voltage-dependent Ca^{2+} channels. When the Ca^{2+} channel is phosphorylated, it is more likely to open upon depolarization. In cardiac muscle cells, Ca^{2+} influx through the Ca^{2+} channels contributes to contraction, and so phosphorylation of Ca^{2+} channels makes the contraction stronger.

Check Your Performance:

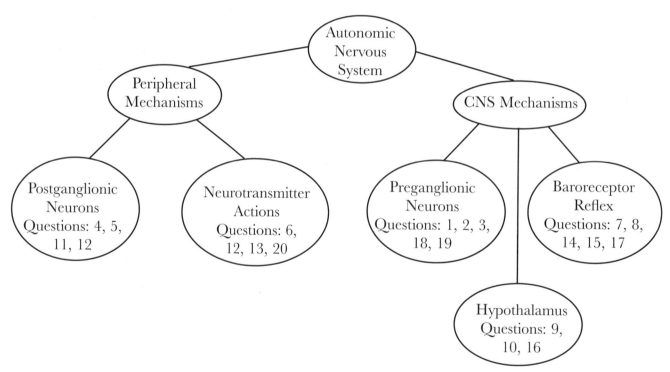

Note the number of questions in each grouping that you got wrong on the chapter test. Identify areas where you need further review and go back to relevant parts of this chapter.

Further help: If you continue to have difficulty with the concepts, a detailed presentation of these topics can be found in chapter 13 of *Neurobiology: Molecules, Cells, and Systems*, published by Blackwell Science, Inc.

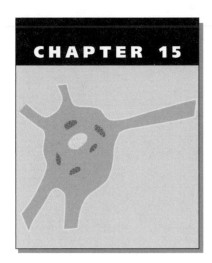

CHAPTER 15

Translating Sensory Information into Motor Signals

Sensory information is important for the proper motor performance. Spinal reflexes are a simple example, but in the brain, the relationship between sensory information and motor output is often more complex. As a model for sensorimotor integration in the brain, this chapter explores the oculomotor system, which combines visual information and vestibular information to control eye movements.

ESSENTIAL BACKGROUND

- Anatomy of the eye and eye muscles
- Sensory transduction in hair cells
- Anatomy of the brain

TOPIC 1: VESTIBULO-OCULAR REFLEX

KEY POINTS

✓ *What is the sensory apparatus of the vestibular system?*

✓ *What parts of the brain stem are important for controlling eye movements?*

✓ *How is the vestibulo-ocular system arranged to produce appropriate reflexive eye movements to compensate for head motion?*

The **vestibulo-ocular reflex** maintains a constant view of external space during rotation and movement of the head. Sensory information to correct for head movements comes from the **semicircular canals**, which form three loops oriented at right angles to each other, one loop for each axis of rotation of the head. A swelling, the **ampulla**, at the end of each loop contains hair cells that respond to motion of the fluid within the semicircular canals. In addition, static position of the head with respect to gravity is signaled by two other parts of the inner ear, the **saccule** and the **utricle**. The cochlea, semicircular canals, saccule, and utricle are collectively called the **labyrinth**. Hair cells of the semicircular canals make synapses onto sensory neurons of the **vestibular ganglion**, located in the inner ear just outside the labyrinth. Axons of the vestibular ganglion form part of cranial nerve VIII, the **auditory nerve** (also called the vestibulocochlear nerve) and terminate in the **vestibular nuclei** in the brain stem. The vestibular nuclei send outputs to the brain stem nuclei containing motor neurons of the eye muscles.

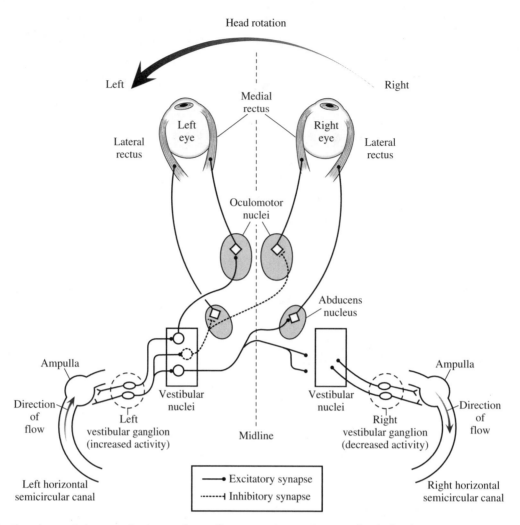

Figure 15.1 Circuitry of the vestibulo-ocular reflex. Head rotation to the left elicits eye movement to the right. Sensory information from the horizontal semicircular canal is distributed to the appropriate motor neurons by the interneurons of the vestibular nuclei.

Each semicircular canal produces eye movements in its own plane. For example, stimulation of the horizontal canal produces horizontal eye movements. Motor neurons of the eye muscles are located in the oculomotor nucleus (cranial nerve III), trochlear nucleus (cranial nerve IV), and abducens nucleus (cranial nerve VI), and the axons of the motor neurons project to the eye muscles via the three cranial nerves of the same name. **Figure 15.1** shows the connections from the horizontal semicircular canal to the horizontal eye muscles, the lateral and medial rectus muscles. Connections for the other semicircular canals follow the same basic scheme as the horizontal canals, but the exact patterns of projections depend on the required direction of reflexive eye movements.

Topic Test 1: Vestibulo-ocular Reflex

True/False

1. The semicircular canals respond to the static position of the head with respect to gravity.

2. The labyrinth is another name for the semicircular canals.

3. The sensory neurons that receive direct synapses from hair cells of the semicircular canals are located in the vestibular nuclei.

4. In the vestibulo-ocular reflex, incoming axons from the vestibular ganglion make direct excitatory connections on excitatory and inhibitory interneurons of the vestibular nuclei.

5. All motor neurons that control eye muscles are located in the oculomotor nucleus (the nucleus of cranial nerve III) in the brain stem.

6. When the head rotates to the left, action potential activity in neurons of the left vestibular ganglion increases and action potential activity in neurons of the right vestibular ganglion decreases.

Multiple Choice

7. During the vestibulo-ocular reflex activated by horizontal rotation of the head to the right,
   a. the eyes rotate horizontally toward the left.
   b. the horizontal semicircular canal on the right side of the head will be affected by the rotation, but the horizontal canal on the left side will not.
   c. all neurons in the vestibular nucleus on the right side of the brain are inhibited, whereas all neurons in the vestibular nucleus on the left side of the brain are excited.
   d. all of the above.
   e. none of the above.

8. The vestibulo-ocular reflex
   a. involves vestibulospinal projections from the vestibular nuclei to the motor neurons of the eye muscles in the spinal cord.
   b. does not involve sensory feedback to control the accuracy of the eye movements.
   c. involves excitation of motor neurons controlling a particular muscle on one side of the brain and simultaneous inhibition of the motor neurons controlling that same muscle on the other side of the brain, in the case of horizontal eye movements.
   d. a, b, and c.
   e. b and c but not a.

9. Neurons of the vestibular nucleus
   a. receive sensory inputs only from the vestibular system.
   b. are located in the brain stem.
   c. make direct synaptic connections with motor neurons that control eye muscles.
   d. a, b, and c.
   e. b and c but not a.

Topic Test 1: Answers

1. **False.** Hair cells of the semicircular canals are stimulated by motion of the head. Each canal is a fluid-filled tube, and hair bundles of hair cells in the ampulla extend into the fluid. When the head rotates, the wall of the tube moves but the fluid in the tube does not immediately move, because of inertia. Thus, the hair bundles are deflected, and the hair cells are excited or inhibited, depending on the direction of deflection (Chapter 9). If rotation continues in the same direction and at the same speed, inertia of the fluid is

overcome and the fluid eventually moves at the same speed as the wall of the canal. Thus, the fluid moves with respect to the wall of the canal only during changes in head position and not when the position of the head is static (unchanging).

2. **False.** The labyrinth is a collective term for the cochlea, semicircular canals, utricle, and saccule.

3. **False.** The vestibular ganglia and the vestibular nuclei are different. The vestibular ganglion is located in the inner ear and contains sensory neurons that receive synapses from vestibular hair cells. The vestibular ganglion neurons send axons into the brain, where they terminate in the vestibular nuclei of the brain stem.

4. **True.** The vestibular nuclei contain both excitatory and inhibitory interneurons, which in turn send connections to sets of motor neurons that move the eyes in particular directions.

5. **False.** Although most eye muscle motor neurons are located in the oculomotor nucleus, other eye muscle motor neurons are in the trochlear nucleus and abducens nucleus. The oculomotor nerve (cranial nerve III) innervates the superior and inferior rectus muscles, the medial rectus muscle, and the inferior oblique muscle. Cranial nerve IV (the trochlear nerve, innervates the superior oblique muscle, and cranial nerve VI (the abducens nerve, innervates the lateral rectus muscle.

6. **True.** Because the semicircular canals on the two sides of the head are mirror images of each other, the direction of fluid flow in the canal is opposite on the right and left sides during head rotation. Thus, the cilia of hair cells on the two sides are deflected in opposite directions. Cilia are arranged such that rotation of the head toward the left depolarizes hair cells in the left horizontal canal and hyperpolarizes hair cells in the right horizontal canal. The hair cells make excitatory synapses onto the neurons of the vestibular ganglia, and so the vestibular ganglion cells will be excited on the left and inhibited on the right.

7. **a.** The vestibulo-ocular reflex moves the eyes in the appropriate direction to compensate for head motion and keep the visual world stable on the retina. When the head rotates horizontally to the right, the eyes move horizontally to the left. Statement b is incorrect because head rotation always affects the semicircular canals on both sides. Statement c is incorrect because rotation of the head alters the amount of excitation from the vestibular ganglion to the vestibular nucleus on each side.

8. **e.** Statement a is incorrect because eye muscle motor neurons are not located in the spinal cord and the reflex does not involve the vestibulospinal tract. Statement b is correct because the vestibulo-ocular reflex is an example of an open-loop reflex, which lacks feedback control of the motor output to the muscle. The accuracy of the vestibulo-ocular reflex in stabilizing images on the retina depends on precision of the excitatory and inhibitory synapses onto neurons of the reflex circuitry. Statement c is also correct because compensation for head rotation in the horizontal direction requires one eye to rotate toward the nose and the other eye to rotate toward the temple.

9. **e.** Despite the name of the nucleus, the vestibular nuclei receive other sensory input besides vestibular information. For example, the visual system also sends information to the same neurons in the vestibular nucleus that receive vestibular information. Visual information is used to control eye movements in response to moving visual stimuli, such as

in the optokinetic reflex. In this reflex, the eyes move to keep a moving object in a stable position on the retina. Thus, the vestibular nucleus is a general machine that computes the excitation and inhibition of the various oculomotor neurons required to produce particular eye movements.

TOPIC 2: BRAIN SYSTEMS CONTROLLING VOLUNTARY EYE MOVEMENTS

KEY POINTS

✓ *What is a saccade?*

✓ *What determines the direction of eye movement during a saccade?*

✓ *What neuronal circuitry in the brain stem programs contractions of the ocular muscles during a saccade?*

✓ *What brain regions control the saccade generator circuits in the brain stem?*

Saccades are fast voluntary eye movements that bring objects of interest into the center of the visual field. Saccades are initiated by **excitatory burst neurons**, which are located in the reticular formation at the level of the pons. Excitatory burst neurons fire a burst of action potentials just before a saccade and excite motor neurons and interneurons of the appropriate ocular muscles. To make a horizontal saccade to the right, excitatory burst neurons are activated that make excitatory synapses on lateral rectus motor neurons in the abducens nucleus of the right side of the brain. In the abducens nucleus, these excitatory burst neurons also activate interneurons that project to the contralateral oculomotor nucleus, where they excite medial rectus motor neurons. This combination pulls the eyes toward the right. Excitatory burst neurons also send axons to posterior parts of the reticular formation, where they excite **inhibitory burst neurons**, which inhibit motor neurons of antagonist eye muscles. A mirror image circuit exists for horizontal saccades to the left, and other analogous circuits drive vertical saccades. By appropriate combinations of the vertical and horizontal circuits, saccades in any direction can be generated.

Excitatory burst neurons are inhibited by **omnidirectional pause neurons**, in the **raphe nucleus** of the brain stem. These neurons fire continuously at a steady rate, providing sustained inhibition that keeps excitatory burst neurons from firing. Omnidirectional pause neurons stop firing just before and during a saccade in any direction (hence the name), removing the inhibitory influence and allowing excitatory burst neurons to trigger the saccade. The selection of which set of excitatory burst neurons will fire during the permissive period is made by inputs to the burst neurons from eye movement command centers, which also provide inhibition that quiets the pause neurons. The eye movement command centers include the **superior colliculus**, which receives sensory information directly from the retina. A spatial map of the visual field is found in the dorsal layers of the superior colliculus, and a spatial map of saccade direction is found in the ventral layers. Electrical stimulation applied at a particular position in the ventral layers produces eye movement toward the point in the visual world represented in the overlying portion of the dorsal layers. Saccade circuits of the superior colliculus are inhibited by the substantia nigra. The **frontal eye fields** in the frontal cortex trigger saccades by activating the superior colliculus and by activating neurons in the caudate nucleus that inhibit the inhibitory neurons of the substantia nigra.

Topic Test 2: Brain Systems Controlling Voluntary Eye Movements

True/False

1. Excitatory burst neurons of the reticular formation trigger saccades.

2. Inhibitory burst neurons ensure relaxation of antagonist muscles that would oppose a saccade.

3. Omnidirectional pause neurons excite the excitatory burst neurons during a saccade.

4. The direction of a saccade is determined by the set of omnidirectional pause neurons activated by higher eye movement command centers.

5. The superior colliculus receives visual inputs in the dorsal layers.

6. The basal ganglia are involved in regulation of eye movements.

7. Cortical neurons in the frontal eye fields send axons to the superior colliculus and to the brain stem saccade generator circuits.

Multiple Choice

8. The following receive direct synaptic inputs from excitatory burst neurons:
   a. motor neurons of eye muscles
   b. inhibitory burst neurons
   c. excitatory interneurons in motor nuclei of cranial nerves
   d. a and b but not c
   e. a, b, and c

Topic Test 2: Answers

1. **True.** Excitatory burst interneurons provide strong excitation to eye muscle motor neurons, both directly and via excitatory interneurons. This strong input stimulates eye motor neurons to fire a rapid burst of action potentials, producing the saccade.

2. **True.** Inhibitory burst neurons are excited by excitatory burst neurons during a saccade. The inhibitory burst neurons inhibit motor neurons and excitatory interneurons of muscles that oppose the desired saccade.

3. **False.** Omnidirectional pause neurons are inhibitory neurons that prevent excitatory burst neurons from firing except when a saccade is triggered.

4. **False.** Omnidirectional pause neurons stop firing during saccades of all directions. The direction of a saccade is determined by the set of excitatory burst neurons activated by eye movement command centers.

5. **True.** The upper layers of the superior colliculus, near the dorsal surface of the brain stem, receive inputs from the visual system, including the retina.

6. **True.** The caudate nucleus, one of the basal ganglia, receives excitatory input from the frontal eye fields of the cerebral cortex. Neurons in the caudate nucleus then inhibit

neurons of the substantia nigra that in turn inhibit the saccade output neurons of the superior colliculus. The caudate nucleus removes the brake (from the substantia nigra) that prevents saccades at the level of the superior colliculus.

7. **True.** Cells of the frontal eye field initiate saccades by exciting the motor output layers of the superior colliculus and the excitatory burst neurons of the brain stem saccade generator.

8. **e.** All targets listed receive direct synapses from the excitatory burst neurons.

IN THE CLINIC

Eye movements involve several parts of the brain, and eye movements are readily observed in routine clinical examination. Therefore, disorders of eye movement are an indicator of several neurological conditions. Defects of the vestibular system can cause abnormal eye movements, such as nystagmus. A normal component of the optokinetic reflex is nystagmus, which is a steady movement in one direction (as when following a moving stimulus) followed by a rapid saccade in the opposite direction to recenter the eyes. Normally, when the head is motionless, the amount of activity in the vestibular ganglion neurons on both sides of the head is the same, and so the eyes remain still. If the semicircular canals or vestibular nerve are damaged on one side, however, reduced activity on that side produces an imbalance that mimics rotation of the head away from the damaged side. The eyes then drift just as if the head were moving. When the eyes reach the extreme position in the orbit, a rapid saccade is executed to recenter the gaze, and the process begins again. Thus, nystagmus when the head is still is one sign of disease of the vestibular pathway.

Chapter Test

True/False

1. The vestibulo-ocular reflex and the saccade generator circuits use the same set of interneurons in the brain stem to connect to eye muscle motor neurons.

2. The vestibulo-ocular reflex and the optokinetic reflex both use neurons of the vestibular nucleus to control eye movements.

3. Rotation of the head is detected by hair cells in the ampulla of each semicircular canal.

4. Neurons of the vestibular nucleus respond to both visual and vestibular stimuli.

5. Neurons of the vestibular ganglia project directly to ocular motor neurons, without any intervening synaptic relay.

6. In the vestibulo-ocular reflex, feedback from stretch receptors of the ocular muscles corrects eye movement.

7. During a saccade, both excitatory burst neurons and inhibitory burst neurons fire a burst of action potentials.

8. The saccade generator circuit of the brain stem does not involve interneurons whose axons cross from one side of the brain to the other.

9. Omnidirectional pause neurons are located in the deep layers of the superior colliculus.

10. The frontal eye fields of the frontal cortex initiate the command to generate a saccade.

Multiple Choice

11. The saccade circuitry involves which of the following brain regions:
    a. raphe nucleus
    b. red nucleus
    c. caudate nucleus
    d. all of the above
    e. a and c but not b

12. To initiate a saccade, the following must normally occur:
    a. omnidirectional pause neurons stop firing
    b. excitatory burst neurons fire
    c. inhibitory neurons of the substantia nigra are inhibited
    d. a and b but not c
    e. a, b, and c

13. During the vestibulo-ocular reflex activated by rotating the head to the left in the horizontal plane,
    a. motor neurons in the oculomotor nucleus on the left side of the brain are excited.
    b. motor neurons in the abducens nucleus on the left side of the brain are excited.
    c. the rate of action potentials decreases in vestibular ganglion neurons on the left side.
    d. all of the above.
    e. none of the above.

14. The semicircular canals
    a. are part of the cochlea.
    b. contain hair cells that synapse onto vestibular ganglion neurons, whose axons enter the brain in cranial nerve VIII.
    c. do not respond initially when the head begins to rotate, because inertia of the fluid inside the canals initially prevents the fluid from moving.
    d. a and b but not c.
    e. b and c but not a.

15. During the vestibulo-ocular reflex activated by rotating the head in the horizontal plane,
    a. excitatory burst neurons in the brain stem reticular formation fire rapid bursts of action potentials to produce saccades in the direction opposite to rotation.
    b. inhibitory burst neurons in the brain stem reticular formation fire rapid bursts of action potentials to inhibit motor neurons of muscles that move the eye in the same direction as rotation.
    c. omnidirectional pause neurons stop firing to allow the reflexive eye movements to occur.
    d. all of the above.
    e. none of the above.

Short Answer

16. List the three cranial nerve nuclei that contain motor neurons of the ocular muscles.

17. Identify the following: raphe nucleus, superior colliculus, and frontal eye fields.

Essays

18. Briefly describe the functional characteristics of the dorsal (upper) and ventral (deeper) layers of the superior colliculus.

19. Beginning with the hair cells, list the synaptic stages leading from the horizontal semicircular canal on the left side of the head to the medial rectus muscle of the right eye.

20. Describe the series of events that normally must occur to produce a saccade.

Chapter Test Answers

1. **F** 2. **T** 3. **T** 4. **T** 5. **F** 6. **F** 7. **T** 8. **F** 9. **F** 10. **T** 11. **e** 12. **e**

13. **a** 14. **b** 15. **e**

16. Oculomotor nerve (cranial nerve III), trochlear nerve (cranial nerve IV), and abducens nerve (cranial nerve VI).

17. Raphe nucleus: a nucleus near the midline of the brain stem at the level of the abducens nucleus that contains the omnidirectional pause neurons of the saccade circuits. Superior colliculus: a combined sensory and motor area on the top surface of the brain stem; the superior colliculus receives visual inputs and provides outputs to the saccade generator circuits in the brain stem that determine direction and magnitude of a saccade. Frontal eye fields: part of the frontal cortex thought to provide the overall command to initiate a saccade, with the actual motor programming carried out in the brain stem.

18. The dorsal layers of the superior colliculus receive sensory inputs from the visual system. The dorsal layers contain a two-dimensional map of visual space, with cells at a particular location responding best to visual stimuli at a particular location in the visual field. The ventral layers represent a two-dimensional map of eye movement related to visual space. Activity of neurons at a particular location in the ventral layers stimulates a saccade that centers the eyes at a particular location in the visual world. The two maps in the dorsal and ventral layers are in point-by-point registration. For example, if cells at a particular location in the dorsal layers respond best to stimuli in the upper right quadrant of the visual field, then activation of cells in the same position in the underlying ventral layers will produce a saccade that moves the fovea toward the upper right. The cells in the superior colliculus somehow translate the visual map into a "saccade map" of motor output for recentering the eyes at different locations in visual space.

19. Hair cells make excitatory synapses onto vestibular ganglion neurons. Vestibular ganglion neurons make excitatory synapses onto inhibitory neurons in the vestibular nucleus on left side of brain stem. The inhibitory vestibular nucleus neurons send axons to the oculomotor nucleus on the right side and make inhibitory synapses onto medial rectus motor neurons. Axons of the motor neurons project to the medial rectus muscle via cranial nerve III and make excitatory synapses onto the muscle fibers.

20. The basic circuitry to produce a saccade is found in the reticular formation of the brain stem. Excitatory burst neurons in the upper part of the reticular formation, at the level of the pons, make excitatory connections with interneurons and motor neurons in the motor nuclei of the eye muscles in the brain stem. Different sets of excitatory burst neurons make connections appropriate for horizontal or vertical saccades to the left or right. Excitatory burst neurons also activate inhibitory burst neurons at lower levels of the

reticular formation, which make appropriate connections to relax eye muscles antagonistic to the direction encoded by the excitatory burst neurons. Excitatory burst neurons are normally prevented from firing by strong inhibitory input from omni-directional pause neurons in the raphe nucleus. Commands for saccades originate in the frontal eye fields of the frontal cortex, which provide excitatory inputs to the caudate nucleus, the saccade programming circuitry of the superior colliculus, and the excitatory burst neurons of the brain stem. The caudate nucleus inhibits neurons in the substantia nigra that make inhibitory synapses in the superior colliculus. Thus, the superior colliculus is activated directly by the cortical command and indirectly by inhibition of an inhibitory input from the substantia nigra. Output from the ventral layers of the superior colliculus inhibits the omnidirectional pause neurons in the raphe nucleus and excites the subset of excitatory burst neurons appropriate for the desired saccade. In addition, direct input from the frontal eye fields to the excitatory burst neurons helps determine which group of excitatory burst neurons fires and thus the saccade direction.

Check Your Performance:

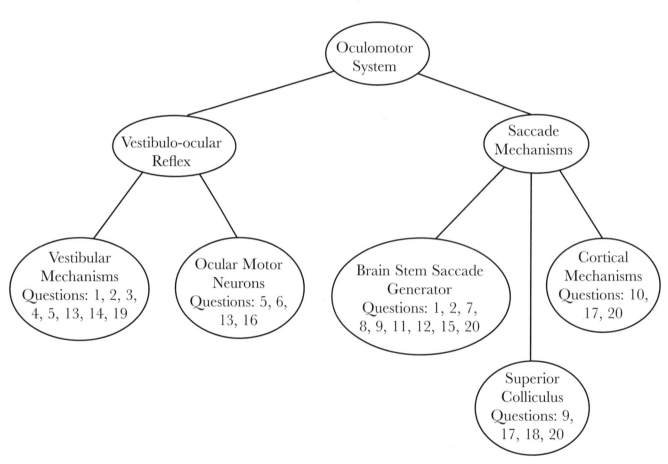

Note the number of questions in each grouping that you got wrong on the chapter test. Identify areas where you need further review and go back to relevant parts of this chapter.

Further help: If you continue to have difficulty with the concepts, a detailed presentation of these topics can be found in chapter 12 of *Neurobiology: Molecules, Cells, and Systems*, published by Blackwell Science, Inc.

UNIT V:
HIGHER NEURAL FUNCTIONS

CHAPTER 16

The Limbic System: Homeostasis, Motivation, and Emotion

The limbic system is a collective term for a group of highly interconnected brain areas that coordinate actions related to homeostasis, sexual behavior, motivation, and emotion. This chapter describes the organization of the limbic system and the function of some of its parts.

ESSENTIAL BACKGROUND

- **General anatomy of the brain**
- **Organization of the autonomic nervous system**

TOPIC 1: ANATOMY OF THE LIMBIC SYSTEM

KEY POINTS

✓ *What parts of the brain form the limbic system?*

✓ *Where is the hypothalamus located?*

✓ *What are the subdivisions of the hypothalamus?*

The core of the limbic system is the **Papez circuit**, a loop consisting of the cingulate gyrus of the neocortex, entorhinal paleocortex, hippocampus, hypothalamus, thalamus, and then back to the cingulate gyrus (**Figure 16.1**). The **cingulate gyrus** runs front to back within the inter-hemispheric fissure separating the two hemispheres at the midline. Axons from the cingulate gyrus to the entorhinal cortex form a fiber bundle called the **cingulum. Entorhinal cortex** is part of the olfactory cortex (Chapter 10), and it sends axons to the hippocampus via the **perforant path**. The **hippocampus** is a C-shaped structure deep within the lower posterior lip of the neocortex, near the border between the cerebrum and the midbrain. Axons from the hippocampus collect in a large fiber bundle called the **fornix**, which has numerous synaptic targets, including the **mammillary bodies** of the hypothalamus. Output axons of the mammillary bodies run in the **mammillothalamic tract** to the **anterior nuclei of the thalamus**, which then send axons to the cingulate gyrus to complete the loop.

In addition to the Papez circuit, other structures associated with the limbic system include orbitofrontal neocortex, parahippocampal gyrus of the neocortex, amygdala, septal area, and

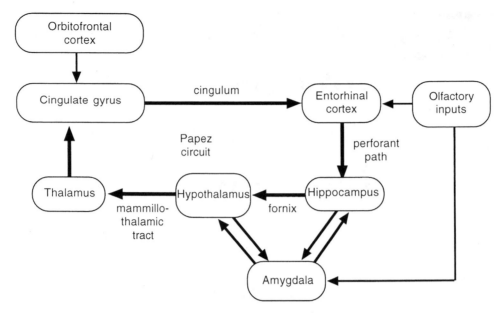

Figure 16.1 The organization of the limbic system. The principal brain regions that make up the core of the limbic system (the Papez circuit) are connected with thick arrows.

parts of the midbrain reticular formation. The **amygdala** is a group of nuclei at the medial tip of the temporal lobe that are interconnected with the hypothalamus and hippocampus. Also, the amygdala is part of the olfactory cortex and thus provides an important route for olfactory information to enter the limbic system. Olfactory information is important for a number of the behavioral functions of the limbic system, such as sexual behavior and feeding.

A central part of the limbic system is the **hypothalamus**, which extends along the base of the diencephalon from the mammillary bodies at the posterior end to the most anterior edge of the diencephalon. Although the subdivisions are indistinct, the hypothalamus is divided into anterior, middle, and posterior regions from front to back and periventricular, medial, and lateral regions from medial to lateral. The periventricular nuclei surround the third ventricle of the brain. The **preoptic area** is anterior and dorsal to the most anterior part of the hypothalamus. Although technically part of the telencephalon, the preoptic area is usually considered to be part of the hypothalamus.

Topic Test 1: Anatomy of the Limbic System

True/False

1. The limbic system includes neocortical, paleocortical, and noncortical regions.

2. The Papez circuit extends from the hypothalamus to the hippocampus and then back to the hypothalamus.

3. The amygdala is part of the olfactory system and part of the limbic system.

4. The fornix is a bundle of axons containing outgoing axons of hippocampal neurons.

5. The hypothalamus is the most ventral portion of the thalamus, surrounding the third ventricle at the base of the telencephalon.

6. Parts of the thalamus are included in the limbic system.

Multiple Choice

7. The Papez circuit includes
   a. the hippocampus.
   b. the entorhinal cortex.
   c. the perforant path.
   d. all of the above.
   e. none of the above.

8. From medial to lateral, the hypothalamus is divided into the following areas:
   a. medial, ventromedial, and anterior.
   b. preoptic, lateral, and mammillary bodies.
   c. periventricular, medial, and lateral.
   d. anterior, middle, and posterior.
   e. none of the above.

9. The preoptic area
   a. is part of the telencephalon but is functionally part of the hypothalamus.
   b. is the part of the thalamus that receives inputs from the mammillary bodies.
   c. is located in the diencephalon.
   d. a, b, but not c.
   e. b and c, but not a.

Topic Test 1: Answers

1. **True.** The limbic system includes a variety of different brain regions that have complex interconnections. Neocortical areas such as the cingulate gyrus are part of the limbic system, as are paleocortical areas such as the olfactory cortex (e.g., entorhinal cortex). Many noncortical structures are also part of the limbic system, such as the hypothalamus.

2. **False.** The Papez circuit involves a more complex set of connections than just the hippocampus and hypothalamus. The Papez circuit is shown in Figure 16.1.

3. **True.** The amygdala is part of the limbic system that makes extensive interconnections with the hypothalamus and hippocampus. Also, the amygdala is considered part of olfactory cortex.

4. **True.** Neurons of the hippocampus project to various targets, and their axons form a large bundle called the fornix. A major target for the fornix is the mammillary bodies of the hypothalamus.

5. **False.** Although the hypothalamus surrounds the third ventricle, it is not part of the thalamus. Also, the hypothalamus is part of the diencephalon, not the telencephalon.

6. **True.** The mammillary bodies of the posterior hypothalamus project to the anterior nuclei of the thalamus via the mammillothalamic tract. The anterior thalamus then sends axons to the cingulate gyrus. Thus, in addition to its roles in sensory and motor systems, the thalamus is also involved in emotion, drives, and homeostasis.

7. **d.** Part of the Papez circuit is the entorhinal cortex, which is connected to the hippocampus via the perforant path.

8. **c.** The hypothalamic nuclei nearest the midline are the periventricular nuclei surrounding the third ventricle. Moving laterally, the next region is the medial hypothalamus and then the lateral region.

9. **a.** The preoptic area is part of the telencephalon, but neurons of the preoptic area make extensive synaptic connections with the hypothalamus, and the preoptic nuclei play functional roles related to limbic system function. Thus, the preoptic area is usually considered part of the hypothalamus.

TOPIC 2: FUNCTIONAL ROLES OF THE HYPOTHALAMUS

KEY POINTS

✓ *What are the parts of the pituitary gland?*

✓ *How does the hypothalamus control the pituitary gland?*

✓ *What parts of the hypothalamus are involved in regulating body temperature?*

✓ *What parts of the hypothalamus are involved in regulating feeding and drinking?*

An important target of the hypothalamus is the **pituitary gland**, which hangs from the bottom of the hypothalamus on a connecting stalk called the infundibulum. The pituitary is a master control gland that is in turn controlled by a subset of neurons in the hypothalamus. These neurosecretory neurons have modified synaptic terminals that act like endocrine cells, releasing hormones into the bloodstream. The pituitary consists of two lobes: the **anterior lobe (adenohypophysis)** and the **posterior lobe (neurohypophysis)**. Axons of **magnocellular (large-cell) neurosecretory cells** of the hypothalamus project directly to the posterior pituitary, where they release the peptide hormones vasopressin or oxytocin into blood vessels for distribution throughout the body. So, the posterior pituitary does not actually manufacture and release hormones. By contrast, the anterior pituitary contains endocrine cells that release seven different hormones into the bloodstream (**Table 16.1**), under the regulation of substances (stimulatory or inhibitory release factors) released by **parvocellular (small-cell) neurosecretory neurons** of the hypothalamus into special **portal capillaries**. The portal vessels transport the release factors to the endocrine cells in the anterior pituitary.

Regulation of body temperature is another function of the hypothalamus. Cold-sensitive and warm-sensitive neurons in the anterior hypothalamus and in the preoptic area respond to cooling and warming of the brain and to changes in skin and body temperature. Electrical stimulation of the anterior hypothalamus produces behavioral (e.g., panting) and autonomic responses (e.g., dilation of skin blood vessels) that dissipate heat. Stimulation of the posterior hypothalamus produces opposite responses (vasoconstriction and shivering) favoring heat production and conservation.

The hypothalamus also regulates drinking, which is stimulated by increased ECF osmolarity or by loss of blood volume, both of which occur if water intake is insufficient. Hypothalamic osmoreceptors detect increased external osmolarity and promote water retention and activate thirst circuitry of the limbic system. Loss of blood volume is signaled to the hypothalamus by both neural and chemical signals. Baroreceptors in blood vessels provide the neural signal, and the chemical signal originates in the kidney, which releases the protein **renin** into the blood-

Table 16.1 Pituitary Hormones, Their Actions, and Hypothalamic Control of Their Release

	RELEASED HORMONE	EFFECT OF RELEASED HORMONE	HYPOTHALAMIC REGULATION OF RELEASED HORMONE
Posterior pituitary hormones	vasopressin	water retention; vasoconstriction	direct: released by magnocellular neurosecretory neurons
	oxytocin	milk ejection from mammary glands; uterine contraction	direct: released by magnocellular neurosecretory neurons
Anterior pituitary hormones	thyrotropin (TSH)	increased thyroxin release from thyroid gland	thyrotropin releasing hormone (+)
	adrenocorticotropic hormone (ACTH)	increased cortisol release from adrenal gland	corticotropin releasing hormone (+)
	growth hormone (somatotropin)	whole body growth	growth hormone; releasing hormone (+); somatostatin (–)
	gonadotropins (LH, FSH)	development of reproductive cells and release of sex hormones from gonads	gonadotropin releasing hormone (+)
	prolactin	milk production by mammary glands	prolactin releasing hormone (+); dopamine (–)
	melanocyte stimulating hormone (MSH)	melanocytes and body coloration changes	MSH releasing factor (+); MSH release inhibiting factor (–)

Abbreviations: TSH, thyroid stimulating hormone; ACTH, adrenocorticotropic hormone; LH, luteinizing hormone; FSH, follicle stimulating hormone; MSH, melanocyte stimulating hormone.

stream when blood volume falls. Renin is a protease that cleaves the plasma protein **angiotensinogen**, releasing a fragment called angiotensin I, which is converted into **angiotensin II**. Angiotensin II is detected by special receptor neurons in the **subfornical organ**, in the wall of the third ventricle. The subfornical organ then sends the thirst signal to neurons of the preoptic area.

Feeding behavior is also regulated by the hypothalamus. The set point of body weight is determined in the hypothalamus. Damage to the ventromedial nucleus of the hypothalamus produces overeating and obesity, whereas damage to the neighboring lateral hypothalamic area produces aphagia (absence of feeding).

Topic Test 2: Functional Roles of the Hypothalamus

True/False

1. Each nucleus in the hypothalamus has a single well-defined function.

2. All hormones released by the pituitary gland are actually synthesized and released by neurosecretory neurons of the hypothalamus.

3. All hypothalamic neurosecretory neurons release substances directly into blood vessels from their nerve endings.

4. All substances released by neurosecretory neurons of the hypothalamus are peptides.

5. Regulation of body temperature involves neurons in the anterior hypothalamus, the preoptic area, and the posterior hypothalamus.

6. Damage to the lateral hypothalamic area produces aphagia.

7. Body weight regulation involves a hormonal signal detected in the hypothalamus.

Multiple Choice

8. Drinking can be stimulated by
   a. hypothalamic osmoreceptors that detect increased osmolarity of the ECF.
   b. baroreceptors in blood vessels that detect decreased blood volume.
   c. angiotensin-sensitive cells of the subfornical organ.
   d. a and b but not c.
   e. a, b, and c.

Topic Test 2: Answers

1. **False.** Each hypothalamic nucleus typically contains neurons involved in various homeostatic processes. For instance, neurosecretory neurons are scattered in several nuclei of the hypothalamus, and some of the same nuclei are also involved in other functions, such as control of body temperature.

2. **False.** Only hormones of the posterior pituitary are synthesized and released directly by hypothalamic neurosecretory neurons. The hormones of the anterior pituitary are synthesized and released by endocrine cells, which in turn are controlled by neurosecretory neurons.

3. **True.** Magnocellular neurosecretory cells have nerve endings in the posterior lobe of the pituitary and secrete substances directly into the systemic circulation for distribution throughout the body. Parvocellular neurosecretory cells secrete regulatory factors into portal blood vessels that carry the factors to the anterior pituitary, where they stimulate or inhibit the release of hormones.

4. **False.** This statement is almost true, because the sole known exception is dopamine, which is released as an inhibitory factor into portal blood vessels by neurosecretory cells that control prolactin release from the anterior pituitary. All other hormones and release factors secreted by neurosecretory neurons are peptides.

5. **True.** The anterior hypothalamus and preoptic area reduce body temperature, and the posterior hypothalamus increases body temperature.

6. **True.** Neurons in the lateral hypothalamic area are thought to be involved in activating the hunger drive. A neuropeptide called orexin is produced by neurons in the lateral hypothalamic area and stimulates feeding. Damage to orexin-containing neurons may explain aphagia produced by lesions in the lateral hypothalamic area.

7. **True.** Fat cells in the body produce a peptide hormone called leptin. Mutation of the gene for leptin produces overeating and obesity, and the receptor for leptin is found in hypothalamic areas that regulate feeding. Leptin may be a satiety signal that inhibits food intake by activating the ventromedial nucleus of the hypothalamus.

8. **e.** All stimuli listed activate thirst circuits in the limbic system.

Pathology of the limbic system gives rise to symptoms ranging from simple, such as fever, to complex, such as emotional disorders. Fever occurs when the temperature set point is increased, so that body temperature is regulated at an elevated level. Pyrogens released into the bloodstream during the immune response to infection act directly on the preoptic area to increase the temperature set point. Perhaps the most notorious clinical use of the limbic system is frontal lobotomy. Destruction of the orbitofrontal cortex or fiber tracts connecting orbitofrontal and cingulate cortex to the rest of the limbic system reduces aggression and makes experimental animals placid. Therefore, the same procedures were performed on psychiatric patients to control emotional disorders. The advent of psychoactive drugs, coupled with concerns about effectiveness and irreversibility of the surgery, put an end to the practice.

Chapter Test

True/False

1. The limbic system is found exclusively in the diencephalon.
2. Prolactin and thyroid-stimulating hormone are released by the anterior lobe of the pituitary.
3. Angiotensin is released by the kidney.
4. Leptin is released by neurons in the ventromedial nucleus of the hypothalamus.
5. The amygdala has reciprocal connections with the hippocampus and hypothalamus.
6. The cingulate gyrus, orbitofrontal cortex, and parahippocampal gyrus are neocortical regions that are part of the limbic system.
7. The mammillary bodies of the hypothalamus secrete hormones that control the mammary glands.
8. Drinking can be initiated by hypothalamic osmoreceptors that respond to increased osmolarity of the extracellular fluid.
9. Hormones of the anterior pituitary are secreted into the blood by endocrine cells, which are in turn controlled by factors released by hypothalamic neurosecretory cells.
10. Renin is a protease that converts angiotensinogen into angiotensin in the circulatory system.

Multiple Choice

11. Which of the following is a part of the hypothalamus?
    a. mammillary bodies
    b. adenohypophysis
    c. hippocampus
    d. all of the above
    e. b and c but not a
12. A lesion in what area of the hypothalamus would produce overeating and obesity?
    a. lateral hypothalamic area

b. preoptic area

c. mammillary body

d. ventromedial nucleus

e. none of the above

13. The limbic system includes

a. the cingulate gyrus.

b. anterior nuclei of the thalamus.

c. entorhinal cortex.

d. all of the above.

e. none of the above.

14. The hormones released in the posterior lobe of the pituitary are

a. prolactin and somatotropin.

b. the gonadotropins luteinizing hormone follicle-stimulating hormone.

c. thyrotropin and adrenocorticotropic hormone.

d. growth hormone and thyroxin.

e. vasopressin and oxytocin.

15. Regulation of body temperature involves neurons located in the

a. preoptic area.

b. anterior hypothalamus.

c. posterior hypothalamus.

d. all of the above.

e. none of the above.

Short Answer

16. Define homeostasis.

17. Identify the following: portal vessels, adenohypophysis, neurohypophysis.

18. Identify the following: gonadotropin, somatotropin, somatostatin.

Essays

19. Describe the series of signals leading from loss of blood volume to activation of thirst circuits in the hypothalamus.

20. Describe the hypothalamic system involved in the control of feeding.

Chapter Test Answers

1. **F** 2. **T** 3. **F** 4. **F** 5. **T** 6. **T** 7. **F** 8. **T** 9. **T** 10. **T** 11. **a** 12. **d**

13. **d** 14. **e** 15. **d**

16. Homeostasis is the process of maintaining a relatively constant internal environment for the cells and tissues of the body. This includes such internal factors as body temperature, energy availability, water content of body fluids, and blood pressure.

17. Portal vessels: local blood vessels that receive release factors secreted by hypothalamic neurosecretory cells and transport them to target cells in the anterior lobe of the pituitary. Adenohypophysis: the anterior lobe of the pituitary gland, which contains endocrine cells that release seven different hormones into the systemic circulation. Neurohypophysis: the posterior lobe of the pituitary, which consists of neural tissue;

hypothalamic magnocellular neurosecretory cells project to the neurohypophysis and secrete vasopressin and oxytocin directly into the systemic circulation.

18. Gonadotropin: collective term for the anterior pituitary hormones luteinizing hormone and follicle-stimulating hormone, which control development of reproductive cells and release of sex hormones in the gonads. Somatotropin: another term for growth hormone, which is secreted by the anterior pituitary. Somatostatin: an inhibitory release factor secreted by parvocellular neurosecretory cells of the hypothalamus to reduce release of growth hormone by the anterior pituitary.

19. Loss of blood volume stimulates water intake by neural and hormonal signals. The neural signal originates from baroreceptors in blood vessels. The hormonal signal is angiotensin II, which is produced in the blood. Reduced blood volume stimulates cells in the kidney to release renin into the bloodstream. In the blood, renin cleaves angiotensinogen, releasing angiotensin I. Angiotensin I is converted to angiotensin II, which is detected by sensory cells in the subfornical organ, located in the third ventricle at the level of the hypothalamus.

20. Control of feeding involves hypothalamic nuclei that stimulate and inhibit feeding. The ventromedial nucleus of the hypothalamus contains neurons that respond to a hormonal satiety signal, leptin, released by fat cells in the body. Activation of this hypothalamic circuit reduces food intake and interference with this signal path (by interfering with production of leptin, by blocking leptin receptors, or by damaging the ventromedial nucleus itself) produces overeating and obesity. The stimulatory circuit for feeding involves the lateral hypothalamic area. The neuropeptide orexin, which is found in neurons of the lateral hypothalamic area, may be an excitatory signal in the hunger circuit. Damage to the lateral hypothalamus produces undereating and weight loss.

Check Your Performance:

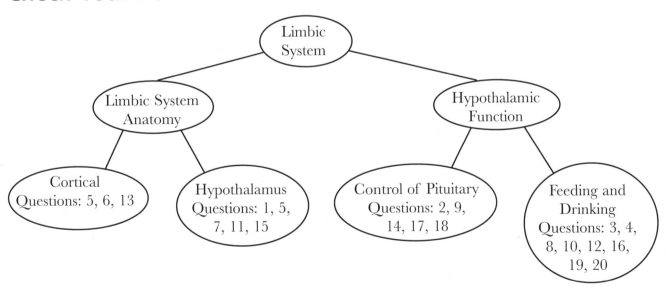

Note the number of questions in each grouping that you got wrong on the chapter test. Identify areas where you need further review and go back to relevant parts of this chapter.

Further help: If you continue to have difficulty with the concepts, a detailed presentation of these topics can be found in chapter 12 of *Neurobiology: Molecules, Cells, and Systems,* published by Blackwell Science, Inc.

CHAPTER 17

Language and Cognition

The chapters concerned with specific sensory systems and with motor systems introduced the cortical regions involved directly in sensory and motor functions. Together, however, primary sensory and primary motor areas account for only about 20% of the human cortex. The remaining areas, sometimes referred to as **association cortex**, are involved in perceptual processing, motor planning, and cognition. **Cognition** is a collective term for complex functions of the brain having to do with language and thought.

ESSENTIAL BACKGROUND

- **General anatomy of the human brain**
- **The lobes of the cerebral cortex**

TOPIC 1: LANGUAGE

KEY POINTS

✓ *What is hemispheric dominance?*

✓ *What is aphasia?*

✓ *What parts of the cerebral cortex are involved in language?*

The two cerebral hemispheres are often not equal in linguistic and cognitive ability. A simple example of hemispheric dominance is hand dominance: over 90% of people are right-handed, meaning that motor circuits of the left cerebral cortex are capable of finer hand and finger control. Almost all right-handers have language function strongly localized to the left hemisphere, and left cortical lesions selectively disrupt language, whereas a lesion in the corresponding part of the right hemisphere does not affect language. Among left-handed people, in whom the right motor cortex produces finer motor control, language function is localized to the left cortex in more than half the population, whereas the right hemisphere is dominant for language in the remaining left-handers. Hemispheric dominance for language emerges slowly during development. A lesion in the left hemisphere of a right-handed child may have no lasting effect on language, because the normally nondominant right hemisphere takes over linguistic functions.

Aphasia is a disorder of speech comprehension or production. Patients with **Broca's aphasia** (or **expressive aphasia**) can understand spoken language but produce speech haltingly, because of a lesion in the lateral frontal cortex of the dominant hemisphere, anterior to the part

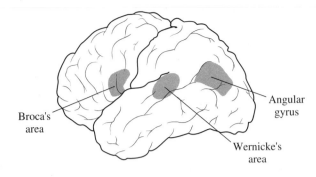

Figure 17.1 Approximate locations of cortical areas important for language in the human brain.

Broca's area

Angular gyrus

Wernicke's area

of primary motor cortex controlling speech muscles. Broca's aphasia is not a simple motor defect, because word pronunciation is normal but smooth flow of words is disrupted. Writing is usually also impaired. **Wernicke's aphasia** (or **receptive aphasia**) involves impaired comprehension of written and oral language, associated with lesions in the temporal lobe surrounding primary auditory cortex. Rhythm of speech production is typically normal, but meaning, word use, and grammar are sometimes disrupted. **Figure 17.1** shows Broca's and Wernicke's areas, where lesions produce the two forms of aphasia.

Topic Test 1: Language

True/False

1. In most people, injection of barbiturate anesthetic into the right carotid artery abolishes speech.

2. In right-handed people, the left cerebral hemisphere is dominant in all cognitive functions.

3. Expressive aphasia is another term for Broca's aphasia.

4. Defective speech production might result from a lesion in the frontal lobe of the right hemisphere.

5. Aphasia could result from a lesion in the fiber tracts (white matter) connecting the frontal cortex with the temporal and parietal lobes.

6. Aphasia is the inability to speak.

Multiple Choice

7. A right-handed patient shows impaired comprehension of speech. Damage to what cortical area would most likely produce this symptom?
   a. frontal cortex
   b. hippocampus
   c. temporal cortex
   d. occipital cortex
   e. paleocortex

8. In Wernicke's aphasia,
   a. comprehension of speech is impaired.
   b. production of speech is normal.

c. a lesion in the frontal lobe would be suspected.
d. all of the above.
e. none of the above.

9. In Broca's aphasia,
   a. pronunciation of individual words is normal.
   b. the ability to write is often also impaired.
   c. comprehension of speech is usually normal.
   d. all of the above.
   e. none of the above.

Topic Test 1: Answers

1. **False.** In most right-handers and half of left-handers, the left cerebral hemisphere is dominant for language. The left carotid artery is the principal blood supply for the left hemisphere, and so injection of anesthetic into the left carotid artery would interfere with language function.

2. **False.** The left hemisphere is dominant for language in almost all right-handers, but the right hemisphere dominates in other cognitive abilities. For example, musical abilities are localized to the nondominant hemisphere (the right hemisphere, for right-handers). Spatial abilities, such as perception of geometrical relationships, are also localized to the nondominant hemisphere.

3. **True.** Patients with Broca's aphasia have impaired speech production.

4. **True.** Broca's aphasia is a defect in speech production resulting from a lesion in the frontal lobe. In most people, the left hemisphere is dominant for language, but in half of left-handers and a few right-handers, language is localized in the right hemisphere. So, a lesion in the right frontal lobe could affect language production in some people.

5. **True.** The arcuate fasciculus is a fiber tract connecting the frontal cortex with more posterior cortical regions, including axons that connect Wernicke's area with Broca's area. Damage to the fasciculus interferes with speech, even if both Wernicke's and Broca's areas are intact. This form of aphasia is called conduction aphasia. Speech comprehension is good in conduction aphasia, but speech production is impaired.

6. **False.** Aphasia refers to defective speech comprehension or production, but many people with aphasia can speak. In fact, some forms of aphasia are associated with excessive speech (logorrhea), although grammar and meaning are disrupted.

7. **c.** Defective speech comprehension is associated with damage in Wernicke's area, surrounding primary auditory cortex in the temporal lobe.

8. **a.** Wernicke's aphasia is impairment of speech comprehension, but production of speech is also impaired. Wernicke's aphasia is caused by lesions in the temporal lobe, not the frontal lobe.

9. **d.** Patients with Broca's aphasia speak only rarely and usually in single words, but they pronounce words normally. Writing is often also impaired, but speech comprehension is good.

TOPIC 2: OTHER COGNITIVE FUNCTIONS

KEY POINTS

✓ *What parts of the cerebral cortex are involved in reading and writing?*

✓ *Damage to which parts of the cortex produce agnosia and apraxia?*

✓ *What changes in personality accompany damage to the frontal lobes?*

Damage to other cortical regions near Wernicke's and Broca's areas produce subtle forms of aphasia. Inability to name things (**anomia**) can result from damage to the posterior temporal lobe, near Wernicke's area (see Figure 17.1). Disruption of reading (**alexia**), sometimes combined with **agraphia** (inability to write), can result from damage in the angular gyrus of the parietal lobe (see Figure 17.1). Damage in the frontal lobe, in front of the primary motor cortex controlling the hand, can also interfere with writing. **Agnosia** is the inability to identify objects using a specific sense, even though primary sensory cortex is intact. Specific forms of agnosia result from damage to cortical areas near primary somatosensory (tactile agnosia), visual (visual agnosia), or auditory (auditory agnosia) cortex. In tactile agnosia, objects cannot be identified by touch alone. Auditory agnosia is the inability to identify nonspeech sounds (e.g., a siren, a barking dog). Inability to recognize colors or faces are examples of visual agnosias. Damage in the language-dominant hemisphere (usually the left hemisphere) causes agnosias involving inability to associate objects with words. Damage to the parietal lobe of the nondominant hemisphere can cause a different form of agnosia, called contralateral neglect, in which the contralateral half of the body is ignored.

Apraxia refers to difficulty with complex series of movements, without paralysis, sensory defects, or impairment of simple tasks. An example is **ideomotor apraxia**, which results from lesions in the **supramarginal gyrus** at the border between the temporal and parietal lobe. In this syndrome, patients can perform each act in a series, but the order is scrambled (e.g., the patient can button buttons, but buttons the shirt before attempting to put it on). Cortical damage can change overall personality. Damage to anterior parts of the frontal lobes interferes with long-term planning and promotes impulsive behavior. Individuals with frontal lobe damage may also be unable to edit speech in socially acceptable ways (e.g., excessive profanity).

Topic Test 2: Other Cognitive Functions

True/False

1. Alexia is the loss of the ability to write.

2. Alexia can occur in isolation, without more general impairment of language abilities.

3. Agnosia refers to the inability to identify objects because of damage to the primary sensory cortex.

4. Agnosia requires cortical damage in the hemisphere that is dominant for language function.

5. A discrete lesion at a specific cortical site produces only a single well-defined deficit in cognitive performance.

6. Apraxia refers to difficulty carrying out sequences of motor actions, without deficit in simple motor acts.

Multiple Choice

7. Damage to the supramarginal gyrus could produce
   a. apraxia.
   b. aphasia.
   c. blindness.
   d. a and b but not c.
   e. a, b, and c.

Topic Test 2: Answers

1. **False.** Alexia is loss of the ability to read. Agraphia is the inability to write.

2. **True.** Although aphasia often involves impaired reading, alexia can occur in isolation, with normal comprehension of oral language.

3. **False.** The term agnosia refers to defects in the ability to identify objects using a particular sense, even though the sensory information is available. A patient with tactile agnosia, for instance, can describe an object placed in the hand but cannot verbally identify it unless it can be seen.

4. **False.** Agnosias associated with the inability to connect objects with language are indeed confined to the language-dominant hemisphere. However, other forms of agnosia can result from damage to the nondominant hemisphere, causing abnormal spatial representation of objects, including the patient's own body.

5. **False.** It is rare that a cortical lesion produces a single isolated defect in cognition. Each region of the cortex is interconnected with many other areas, and neurons at a particular site may participate in various functions. Imaging of activity in human brains during mental tasks reveals that wide-ranging areas of the cortex are active and that a given area is active during more than one task. Also, lesions produced by an accident or a stroke are likely to damage not just the outer cortical layers but also the underlying axons of the white matter, which may connect distant parts of the cortex serving other cognitive functions.

6. **True.** Everyday acts such as getting dressed require that motor programs are carried out in a particular order. In apraxia, cortical damage interferes with the ability to execute the motor programs in the correct order. Each motor program is intact, but the order is incorrect.

7. **d.** The supramarginal gyrus is at the border between the temporal and parietal lobes, where axons of the arcuate fasciculus turn laterally and ventrally to enter Wernicke's area. Thus, lesions in this area often damage the arcuate fasciculus, producing aphasia. Also, the supramarginal gyrus is involved in planning sequences of motor acts. Apraxia is a deficit in the ability to carry out such motor sequences.

Much information about cognitive functions was learned from studies of neurological patients who suffered cortical damage because of head trauma or stroke. Until recently, characterization of the lesion underlying a particular syndrome was possible only postmortem. Development of noninvasive imaging techniques for diagnosis of neurological problems now allows brain lesions to be visualized in living patients. These techniques include computed tomography, a method of constructing three-dimensional images based on X-ray absorption, and magnetic resonance imaging (MRI), which uses the magnetic behavior of the nuclei of molecules to construct a three-dimensional picture of internal structures. These techniques provide useful but static information about the brain structure. Dynamic information about neural activity during cognition is provided by variants of the imaging techniques, such as positron emission tomography (PET) and functional MRI (fMRI). In PET, a subject is injected with a radioactive form of a compound and gamma rays created by positrons emitted by the decaying isotope are used to locate the compound in the brain. To obtain information about neuronal activity, a radioactive glucose analogue (2-deoxyglucose) is injected, which accumulates in active neurons that have higher metabolic need. fMRI detects increased flow of oxygenated blood in more active brain regions and so both PET and fMRI can be used to determine which brain regions are more active when a subject is asked to carry out a particular cognitive task.

Chapter Test

True/False

1. In right-handed people, the left hemisphere is responsible for language, whereas in left-handed people, the right hemisphere is responsible for language.

2. A cortical lesion that causes aphasia would be expected to be located in the left hemisphere for most of the population.

3. Aphasia is a general term for disorders of language function.

4. Wernicke's aphasia is associated with reduced comprehension of spoken language.

5. Broca's aphasia occurs following lesions in the temporal lobe of the language-dominant hemisphere.

6. Disturbance of language function can result only from a lesion in either Broca's area or Wernicke's area and not from lesions in any other areas of the cortex.

7. To produce a deficit in language, lesions must occur in both cerebral hemispheres.

8. Visual agnosia requires that the primary visual cortex in both hemispheres must be damaged.

9. Different forms of agnosia are associated with each of the major senses.

10. Contralateral neglect refers to the inability to detect visual stimuli in the visual field opposite the hemisphere where the lesion is located.

11. A change in personality could result from damage to the frontal lobes.

Multiple Choice

12. After a stroke, a right-handed patient is able to follow verbal commands but produces speech only in single-word utterances. Where is this patient most likely to have cortical damage?
    a. The lateral posterior portion of the frontal lobe on the left side.
    b. The lateral posterior portion of the frontal lobe on the right side.
    c. The middle lateral portion of the temporal lobe on the right side.
    d. The middle lateral portion of the temporal lobe on the left side.
    e. The middle portion of the parietal lobe (the angular gyrus) on both sides.

13. A lesion in the angular gyrus of the parietal lobe would mostly likely be associated with
    a. alexia.
    b. agraphia.
    c. apraxia.
    d. a and b but not c.
    e. a, b, and c.

14. Aphasia is associated with damage to
    a. the lateral portion of frontal cortex in the language-dominant hemisphere.
    b. the temporal cortex surrounding primary auditory cortex in the language-dominant hemisphere.
    c. the arcuate fasciculus in the language-dominant hemisphere.
    d. a and b but not c.
    e. a, b, and c.

15. A neurological patient is observed to put shoes on without socks and then to attempt to put socks on over the shoes. What is your best guess as to where this patient might have a lesion?
    a. postcentral gyrus
    b. precentral gyrus
    c. supramarginal gyrus
    d. angular gyrus
    e. none of the above

Short Answer

16. Identify the following: alexia, agraphia, apraxia.

17. Matching (more than one selection from group B may in some cases be appropriate):
    Group A: alexia, agraphia, aphasia, agnosia, apraxia
    Group B: frontal lobe, parietal lobe, temporal lobe, occipital lobe

18. Identify the following: Wernicke's area, Broca's area.

19. Identify the following: PET, fMRI.

Essay

20. Discuss the following statement: the human brain contains a language center that is responsible for language abilities.

Chapter Test Answers

1. **F** 2. **T** 3. **T** 4. **T** 5. **F** 6. **F** 7. **F** 8. **F** 9. **T** 10. **F** 11. **T** 12. **a**

13. **d** 14. **e** 15. **c**

16. Alexia: a subtype of aphasia characterized by the inability to read, often associated with lesions of the angular gyrus of the language-dominant hemisphere. Agraphia: loss of the ability to write, which is sometimes associated with alexia caused by lesions of the angular gyrus; agraphia can also result from lesions in the frontal lobe, in front of the primary motor area for the hand in the language-dominant hemisphere. Apraxia: inability to perform purposeful movements, without paralysis, difficulty with simple movements, or sensory loss.

17. Alexia: frontal lobe; parietal lobe. Agraphia: frontal lobe; parietal lobe. Aphasia: frontal lobe; parietal lobe; temporal lobe. Agnosia: parietal lobe; temporal lobe. Apraxia: temporal lobe; parietal lobe.

18. Wernicke's area: region in the temporal lobe near the primary auditory cortex. Lesions in this area interfere with the comprehension of language (receptive aphasia), as first described by the neurologist Wernicke. Broca's area: region in the temporal lobe, in front of the part of the primary motor cortex concerned with the muscles involved in speech. Lesions in this area interfere with the production of speech (expressive aphasia), as first described by the neurologist Broca.

19. PET: acronym for positron emission tomography. In this brain imaging technique, a subject is injected with a radioactive form of a compound (such as an analogue of glucose to monitor metabolic activity). Radioactive decay of the compound releases positrons, which in turn generate gamma rays whose source is computed to localize the brain region accumulating the radioactive compound. fMRI: a variant of magnetic resonance imaging that gives information about blood flow in the brain. Blood flow increases when neuronal activity increases, and thus functional MRI gives information about neuronal activity.

20. Like other complex cognitive functions, language requires concerted action of numerous brain regions, and so no single part of the brain could be considered a "language center." Different aspects of language, such as reading, writing, comprehension of spoken language, and expression of spoken language, are more strongly represented in particular cortical regions. Nevertheless, lesions in several different areas affect each of these functions, and a mixture of symptoms results from lesions in a particular locus.

Check Your Performance:

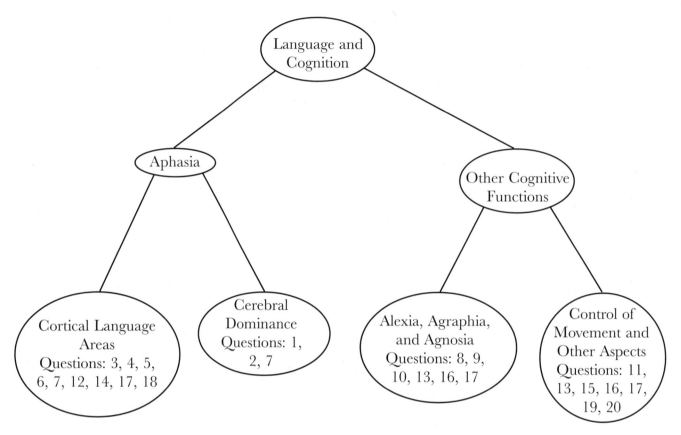

Note the number of questions in each grouping that you got wrong on the chapter test. Identify areas where you need further review and go back to relevant parts of this chapter.

Further help: If you continue to have difficulty with the concepts, a detailed presentation of these topics can be found in chapters 53 and 54 of *Principles of Neural Science, Third Edition*, by Kandel et al.

CHAPTER 18

Learning and Memory

Even animals with simple nervous systems learn to associate neutral stimuli with biologically significant events. Once learned, these behavioral changes are stored in memory for periods of minutes, days, or years. This chapter summarizes what is known about neural mechanisms underlying learning and memory.

ESSENTIAL BACKGROUND

- **Mechanisms of synaptic transmission**
- **The action potential**
- **G proteins, G protein-coupled receptors, and second messengers**
- **Mechanisms of gene transcription**

TOPIC 1: LONG-TERM POTENTIATION IN THE HIPPOCAMPUS

KEY POINTS

✓ *What is the role of the hippocampus in memory?*

✓ *What is long-term potentiation?*

✓ *How do cellular mechanisms of synaptic transmission account for long-term potentiation?*

Change in synaptic strength can alter behavior controlled by a neural circuit. **Long-term potentiation** (**LTP**) is enhancement of synaptic strength lasting hours or days. LTP has been particularly well studied in the **hippocampus**, part of the limbic system involved in the memory. In the hippocampus, a high-frequency burst of action potentials in a synaptic input increases the amplitude of subsequent excitatory postsynaptic potentials evoked by single action potentials. Potentiation of the synaptic connection persists for as long as a week (hence, the name long-term potentiation). Synaptic transmission is enhanced after strong stimulation not only in the synapses that received the strong stimulation but also in synapses that were only weakly stimulated, as long as the weak stimulation was temporally associated with the strong stimulus. Thus, potentiation is heterosynaptic and associative.

LTP is initiated by large depolarization of the postsynaptic neuron. Under normal circumstances, sufficient depolarization occurs only when multiple synaptic inputs are repetitively stimulated. This requirement for strong postsynaptic depolarization explains why high-frequency stimulation of the input fibers is normally required for LTP. Excitatory synapses in the hippocampus use the neurotransmitter glutamate. In at least some instances, LTP is triggered by Ca^{2+} influx through a class of postsynaptic glutamate receptors, called N-methyl-D-aspartate (NMDA) receptors, that require both glutamate and postsynaptic depolarization to allow ions to enter. Increased postsynaptic calcium triggers cellular messengers that enhance future transmission via both presynaptic and postsynaptic changes.

Topic Test 1: Long-term Potentiation in the Hippocampus

True/False

1. Destruction of the hippocampus erases all memories.

2. LTP can be activated by a high-frequency burst of action potentials in excitatory synaptic inputs to hippocampal neurons.

3. Activation of LTP in a hippocampal neuron requires a high-frequency burst of action potentials in synaptic inputs.

4. During LTP, all excitatory synaptic inputs to a hippocampal neuron are equally potentiated.

5. NMDA receptors are a class of glutamate receptor.

6. To conduct ions across the membrane, NMDA receptors must bind glutamate simultaneously with depolarization of the postsynaptic cell.

7. LTP occurs only in the hippocampus.

Multiple Choice

8. LTP in the hippocampus requires
   a. postsynaptic depolarization.
   b. presynaptic action potential activity.
   c. influx of Ca^{2+} into the postsynaptic cell.
   d. all of the above.
   e. none of the above.

9. Increased amplitude of the excitatory postsynaptic potential during LTP occurs because of
   a. increased release of neurotransmitter per action potential in the presynaptic terminal.
   b. increased number of postsynaptic glutamate receptors that open when glutamate is released from the presynaptic terminal.
   c. release of a retrograde signal from the postsynaptic cell that feeds back to affect the presynaptic terminal.
   d. a and b but not c.
   e. a, b, and c.

10. LTP is an increase in synaptic strength lasting
   a. milliseconds to tenths of a second.
   b. seconds to minutes.

c. hours to days.

d. months to years.

e. years to decades.

Topic Test 1: Answers

1. **False.** If the hippocampus is destroyed, short-term memory (memory lasting a few seconds) is unimpaired. Also, long-term memory (memory lasting years) remains intact for events before destruction of the hippocampus. Transfer of new information into long-term memory is severely impaired, however, indicating that the hippocampus is necessary to convert short-term to long-term storage.

2. **True.** Normally, an action potential in a single excitatory synaptic input to a hippocampal neuron elicits a small excitatory postsynaptic potential. After a burst of high-frequency firing, however, a single action potential elicits a larger postsynaptic response.

3. **False.** The crucial aspect of a high-frequency burst of presynaptic firing is that it produces large postsynaptic depolarization. Therefore, LTP can be activated by pairing low-frequency presynaptic firing with postsynaptic depolarization produced by other means.

4. **False.** Only active synaptic inputs are potentiated during LTP. If a synapse is silent during large postsynaptic depolarization, subsequent postsynaptic responses activated by that synaptic input are not enhanced.

5. **True.** NMDA is an acronym for N-methyl-D-aspartate, an artificial glutamate analogue. The natural neurotransmitter recognized by NMDA receptors is glutamate.

6. **True.** Glutamate binding is necessary but not sufficient for ion conduction through NMDA receptors. In addition, the postsynaptic cell must depolarize while glutamate is bound. NMDA receptors do not allow ions to flow in the presence of glutamate at negative membrane potentials because magnesium, a divalent cation, enters the channel from outside when the membrane potential is strongly negative and blocks the channel. If the postsynaptic cell is depolarized when glutamate binds, magnesium is less likely to enter and block the open NMDA receptor.

7. **False.** LTP occurs at many synapses in various brain regions, including the cerebral cortex. However, LTP in the hippocampus is particularly interesting as a possible memory mechanism, given the importance of the hippocampus in aspects of human memory.

8. **d.** LTP is activated only at synapses in which presynaptic action potentials are associated with postsynaptic depolarization. Therefore, statements a and b are correct. Also, Ca^{2+} influx through NMDA receptors in the postsynaptic cell is required for LTP in some parts of the hippocampus.

9. **e.** Evidence suggests that both presynaptic and postsynaptic factors contribute to LTP. Presynaptically, the amount of neurotransmitter released per action potential increases, and there is also an increase in the number of postsynaptic glutamate receptors at the synapse. If LTP has a presynaptic component, then the question arises of how an increase in postsynaptic calcium is signaled to the presynaptic terminal. A retrograde signal is required, which is released from the postsynaptic cell to influence the presynaptic

cell. The identity of the retrograde signal is not clearly established, but one candidate is the gas nitric oxide, which is released from the amino acid arginine by the postsynaptic enzyme nitric oxide synthase.

10. **c.** "Long term" in the case of LTP means hours or days, not months or years. Other briefer forms of potentiation also occur at synapses, often superimposed with LTP. Synaptic facilitation is an enhancement of synaptic strength lasting tens or hundreds of milliseconds. Synaptic augmentation is a longer lasting enhancement that decays within a few seconds.

TOPIC 2: MODIFICATION OF SYNAPTIC STRENGTH IN REFLEX CIRCUITS

KEY POINTS

✓ *What are habituation and sensitization of a reflex?*

✓ *How does synaptic circuitry of a reflex change during habituation and sensitization?*

✓ *What mechanisms produce a long-term shift in strength of a reflex?*

Simple reflex circuits can be modified based on experience. Cellular mechanisms of reflex learning have been studied extensively using the withdrawal reflex in a marine mollusk, *Aplysia californica*. The gill of this sea slug is located in a fleshy pouch at the top of the animal, and the flow of water over the gill is controlled by a muscular spout, the siphon. If the siphon is touched, it and the gill reflexively withdraw into the pouch. **Habituation** of a reflex refers to progressive decline in the reflex with repeated presentations of the eliciting stimulus. In the gill-withdrawal reflex, habituation occurs because excitatory postsynaptic potentials produced in motor neurons by the sensory neurons decrease with repeated stimuli. This depression in the excitatory synapses involves two factors: fewer presynaptic Ca^{2+} channels open during depolarization and the pool of synaptic vesicles available for release decreases in the synaptic terminal.

Sensitization is enhancement of a reflex after contact with a noxious stimulus. Sensitization of the withdrawal reflex in *Aplysia* occurs when facilitatory interneurons are stimulated by the noxious stimulus. **Figure 18.1** shows the neural circuitry involved in sensitization. Facilitatory interneurons release the neurotransmitter serotonin, which acts via G protein-coupled receptors to increase the second messenger, cyclic AMP, in the excitatory synaptic terminals. Cyclic AMP induces phosphorylation of K^+ channels in the presynaptic terminals, and the phosphorylated channels respond less to depolarization. With fewer K^+ channels available to contribute to action potential repolarization, presynaptic action potentials are longer in duration, causing greater Ca^{2+} influx during an action potential and greater neurotransmitter release.

Repetitive bouts of habituation or sensitization produce changes in strength of the withdrawal reflex lasting days or weeks. The transition from short-term to long-term sensitization of the withdrawal reflex requires RNA synthesis and protein synthesis. This suggests that a signal is sent to the cell nucleus after repetitive bouts of sensitization, altering RNA transcription and protein synthesis. The signal for long-term sensitization involves the cyclic AMP signaling pathway, because long-term sensitization can also be triggered by elevated cyclic AMP. Long-term habituation and sensitization also reflect changes in the number of synapses between sensory neurons and motor neurons in the withdrawal reflex. Long-term sensitization is correlated with increased

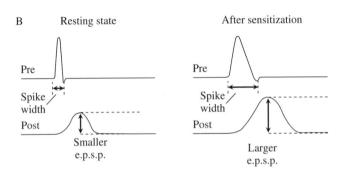

Figure 18.1 Neural basis of sensitization of the gill withdrawal reflex in *Aplysia*. (A) The excitatory synapses in the gill withdrawal circuit receive presynaptic inputs from a facilitatory inter-neuron. The facilitatory interneuron in turn is activated by noxious stimuli applied to the tail of the animal. (B) Sensitization produces larger excitatory postsynaptic potentials in the withdrawal circuit. This occurs because the presynaptic action potential is longer in duration after sensitization.

number of excitatory synapses from the sensory neuron to the motor neuron, whereas the number of synapses declines after long-term habituation.

Topic Test 2: Modification of Synaptic Strength in Reflex Circuits

True/False

1. Sensitization and habituation in the withdrawal reflex of *Aplysia* occur because of changes in the size of excitatory postsynaptic potentials in the motor neurons controlling gill withdrawal.

2. Sensitization of the withdrawal reflex in *Aplysia* can be induced by direct application of serotonin to the presynaptic terminals.

3. Habituation of the withdrawal reflex in *Aplysia* is produced by activation of inhibitory interneurons that counteract excitatory synaptic connections in the reflex circuit.

4. Long-term sensitization of the *Aplysia* withdrawal reflex can be produced by elevation of intracellular cyclic AMP.

5. Blocking protein synthesis prevents both short-term and long-term sensitization of the gill-withdrawal reflex in *Aplysia*.

6. The number of synapses made by a neuron onto another neuron in the nervous system is fixed and unchanging.

7. Learned changes in synaptic strength in reflex circuits occur only in simple organisms like *Aplysia* and are not observed in the mammalian central nervous system.

Multiple Choice

8. Habituation in the gill-withdrawal reflex is caused by
   a. repetitive activation of the reflex.
   b. reduced opening of presynaptic Ca^{2+} channels during presynaptic action potentials.
   c. reduced number of excitatory synapses from sensory neurons onto motor neurons.
   d. a and b but not c.
   e. a, b, and c.

9. Sensitization in the gill-withdrawal reflex is caused by
   a. facilitatory interneurons that release serotonin.
   b. increased number of presynaptic Ca^{2+} channels.
   c. increased opening of presynaptic K^+ channels during an action potential.
   d. a and b but not c.
   e. a, b, and c.

Topic Test 2: Answers

1. **True.** Habituation results from synaptic depression in excitatory synapses in the reflex circuit. The amount of neurotransmitter released by each presynaptic action potential decreases during depression. Sensitization results from increased neurotransmitter release from excitatory synaptic terminals, caused by increased duration of the presynaptic action potential.

2. **True.** Facilitatory interneurons are responsible for sensitization of the withdrawal reflex in response to noxious stimuli. The facilitatory interneuron releases the neurotransmitter serotonin onto excitatory presynaptic terminals in the withdrawal circuit. Thus, direct application of serotonin mimics the effect of the facilitatory interneuron.

3. **False.** Habituation is caused by reduced release of excitatory neurotransmitter, not by activation of an opposing inhibitory input.

4. **True.** Elevation of cyclic AMP in the synaptic terminal underlies short-term sensitization in the withdrawal reflex. Cyclic AMP activates protein kinase A, which then phosphorylates K^+ channels and increases duration of the action potential. In long-term sensitization, activated protein kinase A may move from the synaptic terminal to the cell body, where it phosphorylates and activates transcription factors. The transcription factors activate genes for proteins required to maintain sensitization, such as a proteolytic enzyme that destroys the regulatory subunit of protein kinase A.

5. **False.** Blocking protein synthesis would prevent long-term sensitization but would have no effect on short-term sensitization, which depends on preexisting proteins. The transition from short-term (lasting minutes) to long-term (lasting days or weeks) sensitization does require protein synthesis.

6. **False.** Addition and subtraction of synaptic connections occurs continually in the nervous system. Thus, neural circuits are not fixed but instead are always changing.

7. **False.** Reflex learning occurs in mammals (including humans) and in simple animals. A well-studied example is the vestibulo-ocular reflex (Chapter 15). To function properly, the strength of synaptic interactions in this reflex must produce the exact amount of eye movement to precisely match head movement. Maintenance of a stable visual image by

the reflex is constantly monitored by neural circuits involving the cerebellum, and synaptic strength is updated appropriately to maintain a match between eye movement and head movement. Synaptic strength in the vestibulo-ocular reflex is based on experience and represents a form of learning.

8. **e.** Habituation is activated by repetitive sensory stimuli that evoke the reflex (statement a). Short-term habituation is caused in part by reduced opening of voltage-dependent Ca^{2+} channels in the synaptic terminal (statement b) during depolarization. Long-term habituation involves decreased number of synaptic terminals from sensory neurons to motor neurons (statement c).

9. **a.** Serotonergic facilitatory interneurons are stimulated by noxious stimuli that induce sensitization (statement a). Increased opening of Ca^{2+} channels is not involved in sensitization of the withdrawal reflex in *Aplysia* (statement b). Potassium channels are involved in sensitization, but the channels are less likely to open during depolarization, not more likely (statement c).

IN THE CLINIC

Information about the organization of human memory and differentiation of the different types of memory have come from studies of human neurological patients with defects in particular aspects of memory. For example, memory for motor tasks can be intact even though memory for faces, objects, and places is severely impaired. A role for the hippocampus in human memory was suggested by instances of bilateral damage to the hippocampus in neurological patients. A particularly well-studied case is an individual known by his initials, H.M., whose hippocampus and associated medial portions of the temporal lobe were surgically removed on both sides in an operation to treat epilepsy. After the operation H.M. was unable to remember new information for more than a brief time, although memory for events before the operation was mostly intact. Analysis of H.M.'s deficits gave rise to the idea that the neurons of the hippocampus are involved in the translation of short-term memories into long-term memories.

Chapter Test

True/False

1. In the hippocampus, only synapses that are silent during strong postsynaptic depolarization subsequently exhibit LTP.

2. A retrograde messenger is a signal sent from the postsynaptic cell back to the presynaptic terminal.

3. NMDA receptors conduct calcium ions across the membrane only at hyperpolarized membrane potentials.

4. The hippocampus is the part of the brain where long-term memories are stored.

5. Normally, LTP is triggered when multiple synaptic inputs to a postsynaptic cell are repetitively active, because combined spatial and temporal summation is required to produce strong depolarization of the postsynaptic cell.

6. LTP is thought to involve changes in both postsynaptic and presynaptic events at the potentiated synapse.

7. Habituation of the withdrawal reflex in *Aplysia* occurs because of synaptic depression in the excitatory synapses of the reflex circuit.

8. Serotonin released by facilitatory interneurons produces sensitization of the *Aplysia* withdrawal reflex by directly combining with and opening potassium channels in the presynaptic terminals.

9. In sensitization of the *Aplysia* withdrawal reflex, protein kinase A phosphorylates potassium channels in the synaptic terminal and phosphorylates transcription factors in the cell body of the sensory neurons.

10. Long-term habituation and sensitization involve changes in the number of synaptic terminals of motor neurons that control muscles of the gill and siphon in *Aplysia*.

11. During sensitization of the withdrawal reflex in *Aplysia*, neurotransmitter release from sensory neurons increases because of increased duration of the presynaptic action potential.

12. The strength of excitatory synaptic connections onto ocular motor neurons is modified to ensure that the vestibulo-ocular reflex correctly matches eye movements to head movements.

Multiple Choice

13. Which of the following is not required for LTP in the hippocampus?
    a. at least a small amount of activity in the presynaptic cell
    b. depolarization of the postsynaptic cell paired with activity in the presynaptic cell
    c. a large amount of activity in at least some presynaptic inputs
    d. increased calcium concentration in the postsynaptic cell paired with activity in the presynaptic cell
    e. All of the above are required for LTP.

14. Suppose we record from a neuron in the hippocampus while controlling the activity in three synaptic inputs to the neuron. We stimulate input A at a high frequency for several seconds, producing strong depolarization of the hippocampal neuron. At the same time, we stimulate input B at a low frequency, but input C is silent. An hour later, we test the size of the postsynaptic response produced by single action potentials in A, B, and C. We would find that
    a. input A produces a large (potentiated) epsp, but inputs B and C produce small epsps.
    b. input A produces a small epsp, but inputs B and C produce large (potentiated) epsps.
    c. all three inputs produce large (potentiated) epsps.
    d. inputs A and B produce large (potentiated) epsps, but input C produces a small epsp.
    e. inputs A and C produce small epsps, but input B produces a large (potentiated) epsp.

15. Sensitization of the gill-withdrawal reflex in *Aplysia* could be prevented by
    a. blocking serotonin receptors.
    b. preventing synthesis of intracellular cyclic AMP.
    c. increasing phosphorylation of potassium channels.
    d. a and b but not c.
    e. a, b, and c.

16. Long-term sensitization of the gill-withdrawal reflex in *Aplysia* could be prevented by
    a. blocking protein synthesis.
    b. blocking RNA synthesis.
    c. adding extra copies of the regulatory subunit of protein kinase A inside cells.
    d. a and b but not c.
    e. a, b, and c.

17. The strength of the withdrawal reflex increases during sensitization because
    a. potassium channels are phosphorylated and closed, increasing the duration of the presynaptic action potential.
    b. cyclic AMP concentration decreases in the synaptic terminals.
    c. protein kinase A is destroyed by proteolytic enzymes.
    d. a and b but not c.
    e. a, b, and c.

Short Answer

18. Explain what is meant by the following statement: LTP is associative.

Essays

19. Describe how the properties of the NMDA receptor explain important features of LTP in the hippocampus.

20. Summarize the steps leading to long-term sensitization of the withdrawal reflex in *Aplysia*.

Chapter Test Answers

1. **F** 2. **T** 3. **F** 4. **F** 5. **T** 6. **T** 7. **T** 8. **F** 9. **T** 10. **F** 11. **T** 12. **T**

13. **c** 14. **d** 15. **d** 16. **e** 17. **a**

18. An important feature of learning is that it is associative, which means that it involves learned associations of one stimulus with another or of a behavior with an outcome. LTP is also associative, because a weakly stimulated synaptic input becomes potentiated if its activation is temporally associated with stronger stimulation provided by other inputs. This temporal association is required for potentiation of the weakly stimulated input.

19. NMDA receptors conduct ions across the postsynaptic membrane only when glutamate is bound to the receptor and the postsynaptic cell is depolarized. External magnesium ions enter and block the NMDA channel if the postsynaptic cell is not sufficiently depolarized. This property is responsible for the fact that stimulation of a synaptic input must be paired with strong depolarization to elicit LTP of that synaptic input. Calcium ions enter the postsynaptic cell through the NMDA receptors in the presence of strong postsynaptic depolarization, and the resulting rise in postsynaptic calcium is the signal that initiates LTP.

20. Noxious stimuli activate facilitatory interneurons that release the neurotransmitter serotonin onto excitatory synaptic terminals of the withdrawal-reflex circuit. Serotonin

binds to G protein-coupled serotonin receptors, which then activate internal G proteins that stimulate the synthetic enzyme for cyclic AMP. The resulting increase in cyclic AMP in turn activates the enzyme protein kinase A, which phosphorylates potassium channels in the synaptic terminals. Phosphorylated potassium channels are less likely to open during depolarization and cannot contribute to the repolarizing phase of the action potential, resulting in broadening of the action potential and enhancement of neurotransmitter release from the excitatory synapses. This accounts for short-term sensitization of the withdrawal reflex. With repetitive noxious stimulation, prolonged activation of protein kinase A allows the catalytic subunit of the enzyme to reach the cell body of the neuron, where it phosphorylates and activates transcription factor proteins. The genes activated by the transcription factors are thought to encode proteins that proteolytically destroy regulatory subunits of protein kinase A. This leads to persistent activation of protein kinase A, and thus persistent sensitization of the reflex circuit. Other genes stimulate formation of new excitatory connections in the reflex circuit, which also contributes to persistent increase in the strength of excitatory neurotransmission.

Check Your Performance:

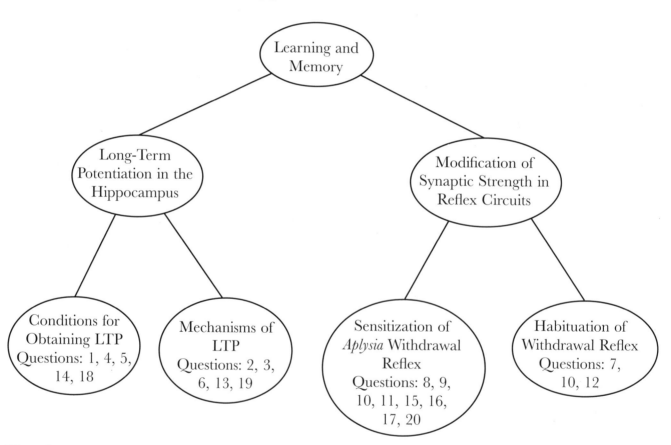

Note the number of questions is each grouping that you got wrong on the chapter test. Identify areas where you need further review and go back to relevant parts of this chapter.

Further help: If you continue to have difficulty with the concepts, a detailed presentation of these topics can be found in chapter 21 of *Neurobiology: Molecules, Cells, and Systems*, published by Blackwell Science, Inc.

UNIT VI:
MOLECULAR NEUROSCIENCE

CHAPTER 19

Molecular Aspects of Neural Development

Development of the nervous system is divided into four phases: neurogenesis, neuronal migration, neurite outgrowth, and synaptogenesis. Molecular signals play important roles in each of these developmental phases, as described in this chapter.

ESSENTIAL BACKGROUND

- **Development of the nervous system (Chapter 2)**
- **Architecture of neurons**
- **Structure of the synapse**
- **Intracellular second messengers**

TOPIC 1: CELL ADHESION MOLECULES AND NEURITE OUTGROWTH

KEY POINTS

✓ *How do migrating cells and growing neurites navigate to find the proper path?*

✓ *What are cell adhesion molecules?*

✓ *What are chemotropic molecules?*

The adult nervous system consists of many different kinds of neurons, each with a particular functional role to play. The spinal cord, for example, contains motor neurons, excitatory interneurons, inhibitory interneurons, sensory projection neurons, and so on, each of which has characteristic synaptic connections and projection patterns. The different neuron types are generated in a fixed sequence, under the control of different regulatory genes that switch on and off in a programmed spatial and temporal sequence. In mammals, for example, homologs of a *Drosophila* gene called *Engrailed* are thought to control the development and differentiation of the cerebellum. As neurons of a particular type mature, neuronal genes are activated in sequence, specifying the characteristics of that cell type in increasingly finer detail.

Migrating neurons and growing neurites must adhere to surrounding cells and the extracellular matrix as they find their way in the developing nervous system. Adhesion is mediated by **cell adhesion molecules**, which are complementary molecules—one on the surface of the neuron and one on the cells or extracellular matrix over which the neuron is traveling—that interact strongly with each other. A wide variety of cell adhesion molecules exist, but a given neuron expresses only a particular subset during neurite outgrowth and cell migration. Thus, neurons can interact with and grow upon some substrates but not others, depending on whether the substrate incorporates adhesion molecules complementary to the neuron's.

Movements of growth cones are also guided by signal molecules that are released into the extracellular space and act at a distance from their source. Such signal molecules either attract growth cones (chemoattractants), causing the neurite to turn toward the source of the signal, or repel growth cones (chemorepellents), causing the neurite to turn away from the source. As a group, chemoattractants and chemorepellents are known as **chemotropic molecules**. Cells that release chemotropic signals serve as guideposts or beacons to aid the navigation of nearby growth cones.

Topic Test 1: Cell Adhesion Molecules and Neurite Outgrowth

True/False

1. The extracellular matrix has the same composition in all parts of the body.

2. Cellular adhesion molecules are molecules on the external surface of the plasma membrane that enable cells to adhere to one another or to the extracellular matrix.

3. All different types of neurons in the nervous system express the same types of cellular adhesion molecules.

4. Cellular adhesion molecules are structural proteins whose only function is to cause cells to stick to each other or to the extracellular matrix.

5. A particular type of chemotropic molecule might attract growth cones of one type of neuron, have no effect on growth cones of a second type, and repel growth cones of a third class of neuron.

6. The expression of genes for a particular type of neuron is genetically preprogrammed when the precursor cell of the neuron is born and cannot be altered by the environment.

Multiple Choice

7. Cellular adhesion molecules can bind to
   a. adhesion molecules of the extracellular matrix.
   b. adhesion molecules on neighboring cells.
   c. chemotropic molecules in the extracellular space.
   d. a and b but not c.
   e. a, b, and c.

8. Which of the following is not a component of the extracellular matrix?
   a. collagen
   b. polysaccharides
   c. glycoproteins
   d. integrins
   e. All of the above are components of the extracellular matrix.

Topic Test 1: Answers

1. **False.** The molecular makeup of the extracellular matrix is not constant in all parts of the organism. Two classes of macromolecules provide the structural backbone of the extracellular matrix: collagen, which is a protein, and glycosaminoglycans, which are large polysaccharide molecules that often combine with proteins to form proteoglycans. The extracellular matrix also contains glycoproteins, which interact with membrane adhesion molecules of growing nerve fibers. Two important glycoproteins of the extracellular matrix are fibronectin and laminin, both of which promote adhesion of cells to the matrix. Tenascin, another glycoprotein found in the extracellular matrix, can inhibit or promote neurite attachment and growth, depending on the type of neuron. The polysaccharide and glycoprotein composition of the extracellular matrix gives a distinct chemical "flavor" to the matrix in the various tissues, which can promote or retard the growth of neurites from different subsets of neurons and thus help to guide neurites to their proper targets.

2. **True.** Four families of cellular adhesion molecules have been identified: integrins, cadherins, selectins, and the immunoglobulin superfamily. Each family consists of a number of related proteins. For example, integrins are formed by the combination of two protein subunits, an alpha subunit and a beta subunit. Several types of both alpha and beta subunits exist, which can associate in various combinations to form more than 20 different types of integrins. Each type of integrin binds particular complementary adhesion molecules in the extracellular matrix, including fibronectin and laminin.

3. **False.** Each neuron subtype expresses selected cellular adhesion molecules. Thus, growing neurites of each neuron type have specific adhesion molecules, which in turn govern the pathways the neurite is able follow as it grows.

4. **False.** In addition to promoting adherence of cells to each other or to the extracellular environment, adhesion molecules also generate intracellular signals based on their interactions with the environment. Cellular adhesion molecules are transmembrane proteins that have an intracellular domain that interacts with internal proteins to generate signals, which can influence the rate of neurite growth. Also, signals generated by cellular adhesion molecules can affect expression of other genes in the neuron, so that the phenotype of the neuron reflects the extracellular environment and intrinsic genetic regulatory programs.

5. **True.** An example of this principle is the chemotropic protein, semaphorin III, which repels growth cones of thermoreceptor and nociceptor sensory neurons of the dorsal root ganglia. Growth cones of mechanoreceptor sensory neurons are unaffected by semaphorin III. This differential action of semaphorin III influences the distribution of axons of these different types of sensory neurons within the spinal cord. Axons of nociceptor, thermo-receptor, and mechanoreceptor neurons enter the spinal cord through the dorsal roots, but nociceptor and thermoreceptor neurons terminate on interneurons in the dorsal part of the gray matter in the spinal cord, and mechanoreceptors continue into the ventral gray matter to make synapses onto motor neurons. At the stage when sensory axons are growing into the spinal cord, cells in the ventral part of the spinal cord secrete semaphorin III, which may in part explain why axons of thermoreceptors and nociceptors do not enter the ventral spinal cord.

6. **False.** Environment affects the differentiation of neuronal precursor cells into particular types of neurons. The parasympathetic and sympathetic portions of the neural crest can be cut out of a developing embryo and transplanted into a recipient embryo with their positions reversed. Normally, the anterior neural crest gives rise to parasympathetic neurons and the middle portion gives rise to sympathetic neurons. After their positions are reversed, sympathetic precursors now give rise to parasympathetic neurons, whereas parasympathetic precursors give rise to sympathetic neurons. Thus, the environment of the precursor cells—not just their genetic programming—determines which set of genes is actually expressed in the adult neurons.

7. **d.** Cell-surface adhesion molecules interact with binding partners either in the extracellular matrix or on the surface of neighboring cells. The integrins, for example, combine with matrix proteins such as laminin and fibronectin. Members of the immunoglobulin superfamily of cellular adhesion molecules, such as neuronal cell adhesion molecule (NCAM) and neuronal-glial cell adhesion molecule (Ng-CAM), are involved in cell-to-cell interactions rather than cell-to-matrix interactions. Chemotropic molecules, however, interact with their own set of receptors in target cells. These receptors are not cell adhesion molecules.

8. **d.** Integrins are one of the major classes of cellular adhesion molecules. All other choices are in fact substances found in the extracellular matrix.

TOPIC 2: NEUROTROPHIC FACTORS AND SYNAPSE FORMATION

KEY POINTS

✓ *What are neurotrophins?*

✓ *How do neurons respond to growth factors?*

✓ *What mechanisms lead to the formation of a synapse?*

Certain classes of secreted proteins, called **neurotrophins**, are required for initiation of neurite outgrowth and for neuronal survival. Because they promote growth and survival, neurotrophins are also called **growth factors**. The best-studied neurotrophin is **nerve growth factor** (NGF), which stimulates neurite outgrowth and is required for survival of dorsal root ganglion and sympathetic ganglion neurons. In addition, NGF acts as a chemotropic signal for growth cones, which are attracted to a source of NGF.

Two neurotrophins closely related to NGF are **brain-derived neurotrophic factor** (BDNF) and **neurotrophin-3** (NT-3). Different neurons have different sensitivities to the three members of the NGF family. Other neurotrophic proteins, such as **ciliary neurotrophic factor**, are not related to the NGF family. In some instances, growth factors that were first discovered to affect the growth of nonneuronal cells also act as neurotrophins. For instance, **fibroblast growth factor** induces proliferation of fibroblasts but also promotes survival of neurons. Nonprotein substances can also act as neurotrophic factors, such as steroid hormones.

Neurotrophic proteins interact with specific receptor molecules that detect particular neurotrophins. For the NGF family of growth factors, the receptor molecules are members of a family of related proteins called **receptor tyrosine kinases** (trk). NGF binds to trk subtype **trkA**,

BDNF binds to **trkB**, and NT-3 binds to **trkC**. NGF is a combination of two identical protein subunits, bound together to form a dimer. Each NGF subunit in the dimer binds separately to a single trkA receptor molecule, bringing the two trkA molecules into close proximity (**Figure 19.1**). The intracellular part of trkA includes a tyrosine protein kinase domain, which phosphorylates tyrosine residues of other protein molecules. When two trkA receptors are brought together by binding of the NGF dimer, each kinase domain phosphorylates the other trkA receptor of the pair. The phosphorylated trkA receptor then becomes activated and initiates signal cascades that produce diverse cellular responses, such as neurite outgrowth and suppression of cell death.

After a growing axon reaches its target, movements of the growth cone stop, and the tip of the neurite converts from a growth cone to a synaptic terminal. Synapse formation has been particularly well studied at the neuromuscular junction. Contact of the motor neuron growth cone with the immature muscle cell induces clustering of acetylcholine receptors in the muscle cell at the point of contact. In addition, nerve contact increases the rate of acetylcholine receptor synthesis in the muscle cell. These effects are mediated by two signal proteins, **agrin** and **acetylcholine receptor-inducing activity (ARIA)**, that are secreted by the nerve terminal into the developing synaptic cleft. Agrin incorporates into the extracellular matrix and helps anchor acetylcholine

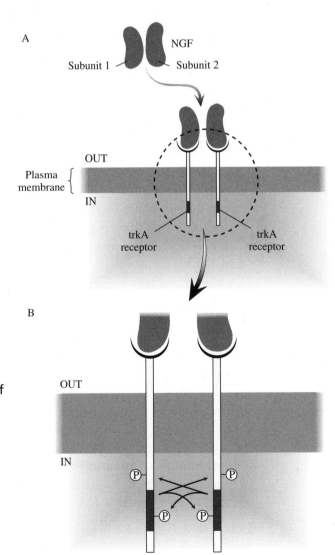

Figure 19.1 Schematic diagram of the interaction of NGF with its receptor molecule. (A) NGF consists of two identical protein subunits, each of which binds to a molecule of the NGF receptor, trkA. (B) The trkA receptors have an intracellular portion that acts as a tyrosine kinase enzyme. The kinase attaches phosphate groups to tyrosine in proteins, including trkA itself. When the two trkA receptors are brought together by NGF binding, they phosphorylate each other. The phosphorylated receptor is then in the activated form and can trigger other signal cascades.

receptors at the appropriate location. ARIA increases expression of the genes encoding the subunits of the acetylcholine receptor subunits. The muscle cell secretes a special form of laminin (**s-laminin**) into the cleft, which acts as a stop signal for motor neuron growth cones.

Topic Test 2: Neurotrophic Factors and Synapse Formation

True/False

1. NGF, BDNF, and NT-3 are three examples of neurotrophins.

2. All neurotrophic factors are proteins.

3. The NGF receptor is a type of G protein-coupled receptor similar to neurotransmitter receptors that activate G proteins inside the cell.

4. Neurotrophins of the NGF family are detected by membrane receptors called receptor tyrosine kinase.

5. The two subunits making up a single molecule of NGF bind to a single trk receptor molecule.

6. Immature muscle cells insert acetylcholine molecules at high density at a single site that will become the neuromuscular junction before the growth cone of the motor neuron arrives during embryonic development.

7. Agrin is an extracellular molecule that is secreted into the synaptic cleft at the neuromuscular junction by the synaptic terminal of the motor neuron.

Multiple Choice

8. At the neuromuscular junction, acetylcholine receptors cluster at the site of synaptic contact because of
   a. interaction of the receptor protein with the cytoskeleton of the muscle cell.
   b. interaction of the receptor protein with the extracellular matrix in the synaptic cleft.
   c. interaction of the receptor protein with membrane proteins of the synaptic terminal of the motor neuron.
   d. a and b but not c.
   e. a, b, and c.

Topic Test 2: Answers

1. **True.** NGF, BDNF, and NT-3 are three members of the NGF family of protein neurotrophins.

2. **False.** It is true that neurotrophins are indeed proteins. However, a neurotrophic factor is simply a substance that promotes the growth and survival of neurons. Steroid hormones are an example of nonprotein hormones that affect neuronal growth and differentiation, such as the effect of sex steroid hormones on parts of the nervous system concerned with sex organs and sexual behavior.

3. **False.** The NGF receptor is trkA, which is a receptor tyrosine kinase that acts by phosphorylating target proteins inside the cell.

4. **True.** Receptor tyrosine kinases are transmembrane proteins whose extracellular portions form the binding site for neurotrophins of the NGF family.

5. **False.** Each of the two subunits making up a molecule of NGF binds to a separate trk molecule in the membrane of the target cell. The binding of the two connected subunits to a receptor molecule brings the two receptors into close proximity, which allows their intracellular domains to interact.

6. **False.** Before the arrival of growth cones from the motor neurons, immature muscle cells insert acetylcholine receptors evenly and at low density throughout the surface of the cell. When a growth cone of a motor neuron contacts the muscle, the acetylcholine receptors begin to cluster at the site of contact.

7. **True.** Agrin is a protein that is manufactured by the motor neuron and secreted into the synaptic cleft, where it is incorporated into the extracellular matrix. Agrin then interacts with a membrane protein called dystroglycan in the plasma membrane of the postsynaptic muscle cell.

8. **d.** Clustering of acetylcholine receptors at the postsynaptic site is maintained by interactions with both the cytoskeleton of the muscle cell and the extracellular matrix of the synaptic cleft. Statements a and b are therefore both correct. However, there is no connection between the receptor and membrane proteins of the synaptic terminal (statement c). The interaction between the acetylcholine receptor and the cytoskeleton is not direct but is instead mediated via a linking protein called rapsyn. Rapsyn interacts with dystrophin, which is attached to the cytoskeleton and to dystroglycan. Dystroglycan is the muscle receptor protein for agrin, the extracellular matrix protein secreted by the nerve terminal. Therefore, the complex formed by agrin, dystroglycan, dystrophin, and rapsyn serves to anchor the acetylcholine receptor to the cytoskeleton of the muscle cell on the one hand and to the extracellular matrix at the synapse on the other hand.

IN THE CLINIC

Damage to the adult nervous system is usually irreversible, in part because adult neurons do not reproduce for the most part. For this reason, there is currently great interest in neurotrophic factors such as BDNF as a means to induce growth and regeneration of damaged neural pathways. Knowledge of neuronal growth factors and their cellular actions is of potentially great benefit in restoration of neural function. Nevertheless, molecular studies of the development of the nervous system teach us that successful restoration of function requires more than just the generation of new neurons and the induction of neurite outgrowth. During development, neurites travel great distances in the nervous system and connect with diverse targets. Successful neural function requires that these connections are reproduced after damage to the intervening tissue. During development, growing neurites require a variety of signposts and beacons that guide them along the correct path. Many of these molecular guides are expressed only transiently during embryonic development and are absent in adults. Restoration of a functional neural circuit, then, requires recapitulation of a complex developmental sequence.

Chapter Test

True/False

1. Proteoglycans are a component of the extracellular matrix.

2. The extracellular matrix consists entirely of protein molecules.

3. A single type of cellular adhesion molecule is expressed by all cells in the nervous system.

4. Neurite guidance involves both cell adhesion molecules and chemotropic molecules.

5. Semaphorin III is an example of a chemotropic molecule.

6. A given chemotropic molecule always acts as either a chemoattractant for all growth cones or a chemorepellent for all growth cones.

7. NGF is a dimer of two identical subunits.

8. NGF, BDNF, and ciliary neurotrophic factor are the only neurotrophins.

9. The NGF receptor is an example of a receptor tyrosine kinase, which phosphorylates tyrosine residues of target proteins.

10. Clustering of acetylcholine receptors at the neuromuscular junction is induced by arrival of the motor neuron growth cone.

11. Agrin is an intracellular molecule in muscle cells that connects acetylcholine receptor proteins to the cytoskeleton.

12. The production of acetylcholine receptor subunits in the muscle cell is increased by a protein factor, ARIA, released from the synaptic terminal of the motor neuron.

Multiple Choice

13. The three members of the NGF family are
    a. NGF, BDNF, and NT-3.
    b. NGF, fibroblast growth factor, and ciliary neurotrophic factor.
    c. NGF, trk, and steroid hormones.
    d. NGF, ARIA, and agrin.
    e. trkA, trkB, and trkC.

14. Which of the following is not a component of the extracellular matrix?
    a. collagen
    b. glycosaminoglycans
    c. laminin
    d. integrin
    e. All of the above are components of the extracellular matrix.

15. The four families of cellular adhesion molecules are
    a. laminins, fibronectins, proteoglycans, and glycosaminoglycans.
    b. NCAM, Ng-CAM, collagen, and laminin.
    c. integrins, cadherins, selectins, and immunoglobulin superfamily.
    d. integrins, NGF, neurotrophins, and laminins.
    e. cadherins, collagen, proteoglycans, and neurotrophins.

16. At the neuromuscular junction,
    a. substances released from the nerve terminal influence the postsynaptic muscle cell.
    b. substances released from the muscle cell influence the nerve terminal.
    c. the extracellular matrix does not extend into the synaptic cleft.
    d. a and b but not c.
    e. a and c but not b.

17. The membrane receptor for NGF
    a. is activated when two receptors are brought in close proximity by binding of the NGF dimer.
    b. is a receptor tyrosine kinase.
    c. is activated when two neighboring receptor molecules phosphorylate each other.
    d. a and b but not c.
    e. a, b, and c.

18. Neurite navigation involves
    a. interaction between cellular adhesion molecules of the neurite and molecules of the extracellular matrix.
    b. interaction between cellular adhesion molecules of the neurite and cellular adhesion molecules of neighboring cells.
    c. chemotropic molecules that attract or repel growing neurites.
    d. a and c but not b.
    e. a, b, and c.

Short Answer

19. Identify the following: laminin, integrin, agrin.

Essay

20. Describe the mechanisms of clustering of acetylcholine receptors in the postsynaptic membrane at the neuromuscular junction.

Chapter Test Answers

1. **T** 2. **F** 3. **F** 4. **T** 5. **T** 6. **F** 7. **T** 8. **F** 9. **T** 10. **T** 11. **F** 12. **T**

13. **a** 14. **d** 15. **c** 16. **d** 17. **e** 18. **e**

19. Laminin: a glycoprotein of the extracellular matrix that interacts with cellular adhesion molecules in the membranes of cells and promotes adhesion of the cell to the matrix. Integrin: a class of cellular adhesion molecule; a molecule of integrin consists of two separate protein subunits, which can be combined in many different combinations, each complementary to particular adhesion molecules in the extracellular matrix. Agrin: a component of the extracellular matrix in the synaptic cleft of the neuromuscular junction; the agrin protein is produced by the motor neuron and secreted into the extracellular space from the synaptic terminal.

20. Clustering of acetylcholine receptors requires that the receptor proteins must be tethered in place in the postsynaptic membrane opposite the synaptic terminal of the motor

neuron. To prevent the receptors from diffusing away, they are attached to the cytoskeleton of the muscle cell by a rapsyn, which is a protein that binds to the complex formed by dystrophin and dystroglycan. Dystrophin connects the cytoskeleton to rapsyn and dystroglycan, which is a transmembrane protein whose extracellular domain interacts with agrin. Agrin is secreted by the synaptic terminal of the motor neuron and becomes incorporated into the extracellular matrix in the synaptic cleft. Therefore, the dystroglycan/dystrophin complex is selectively concentrated at the point of synaptic contact because the extracellular matrix contains agrin only at the synapse. The acetylcholine receptors are then attached to the dystroglycan/dystrophin complex via rapsyn.

Check Your Performance:

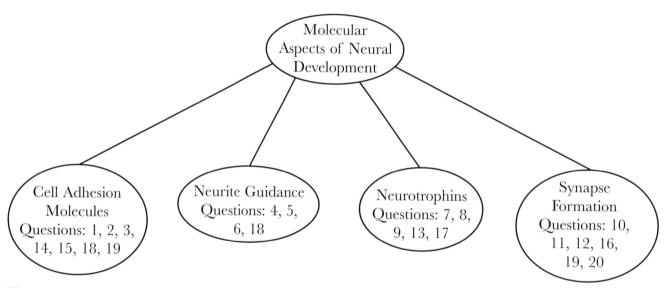

Note the number of questions in each grouping that you got wrong on the chapter test. Identify areas where you need further review and go back to relevant parts of this chapter.

Further help: If you continue to have difficulty with the concepts, a detailed presentation of these topics can be found in chapter 20 of *Neurobiology: Molecules, Cells, and Systems*, published by Blackwell Science, Inc.

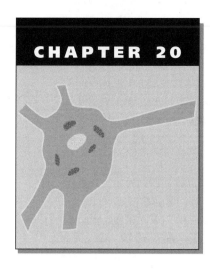

Genetic and Molecular Biological Approaches to the Nervous System

Analysis of the nervous system at the molecular level provides a different approach to how individual neurons function, and how circuits of neurons work together to produce behavior. As with other fields of biology, recent advances in molecular biological techniques allow precise manipulation of the nervous system. This chapter discusses some ways molecular approaches are used in neurobiology.

ESSENTIAL BACKGROUND

- **Mechanisms of gene transcription**
- **Molecular aspects of neural development**
- **Mechanisms of protein translation**

TOPIC 1: GENETIC AND TRANSGENIC APPROACHES

KEY POINTS

✓ *What are some fruitfly mutants that affect the nervous system?*

✓ *What is a transgenic animal?*

The fruitfly *Drosophila melanogaster* is a standard animal for genetic analysis. Many neural genes were first characterized in *Drosophila*. For example, molecular studies of voltage-sensitive potassium channels were first carried out using a *Drosophila* mutant called *shaker*, in which the mutation abolishes expression of a voltage-dependent potassium channel important for repolarization of the action potential. Other interesting *Drosophila* mutations include *shibire*, which is paralyzed because of synaptic vesicle depletion at synapses. The protein encoded by the *shibire* gene is a GTPase called **dynamin** that plays an important role in the recovery of vesicle membrane after neurotransmitter release at the synapse. Depletion of vesicles in the mutant flies occurs because vesicle recycling is disrupted when dynamin cannot function properly.

Mutations of single genes can also affect behavior more subtly. *Drosophila* mutants called *dunce* and *rutabaga*, for example, have impaired associative learning. These mutations affect the cyclic AMP signaling system in the nervous system, which suggests that cyclic AMP is involved in associative learning. Single-gene mutations, including *timeless* and *period*, that affect circadian rhythms have also been identified in *Drosophila*.

In mammals, information about the nervous system has also been obtained from studies of mutants. An example of a neurological mutation in mice is *reeler*, which is characterized by abnormal cell migration during development of the brain. The result is that layered structures such as the cerebral cortex, cerebellum, hippocampus, and olfactory bulbs do not form properly. The mutated gene in the reeler mouse encodes a protein called **reelin**, which is a molecule of the extracellular matrix that serves as a signaling and/or adhesion molecule to guide migrating neurons to their proper positions during early stages of brain development.

Although mutations that occur randomly (whether spontaneously or induced by mutagenesis) have been valuable in establishing the function of proteins encoded by single genes, new techniques that allow targeted disruption of single genes promise to be even more valuable. The ability to insert novel genes allows the production of **transgenic animals**, most commonly mice, in which DNA is introduced artificially into the natural genome. The inserted gene may either add a new gene (e.g., from a different organism) or may replace the natural gene. If the natural gene is replaced by a nonfunctional version, the protein encoded by the gene is lost, or "knocked out." This type of transgenic mouse is therefore called a **knockout** mouse.

Topic Test 1: Genetic and Transgenic Approaches

True/False

1. Mutation of a single gene has no effect on behavior because the behavior of the organism is a polygenic trait.

2. *shaker* and *shibire* are the names of two *Drosophila* strains in which a gene mutation affects nervous system function.

3. To identify the gene encoding a protein, it is first necessary to isolate the protein and determine its amino acid sequence.

4. A transgenic organism is one in which nonnative DNA has been incorporated into the genome of the organism.

5. In a knockout mouse, the function of a protein is tested by replacing the gene that encodes the protein with an altered gene that does not produce a functional protein.

Multiple Choice

6. *Drosophila* melanogaster is an important organism for molecular studies of the nervous system because
   a. genetic analysis more readily allows rapid identification of the chromosomal localization of mutations that affect behavior in *Drosophila*.
   b. *Drosophila* has a larger genome than mammals, which means that more genetic markers exist for *Drosophila*.
   c. transgenic animals can be produced only in *Drosophila*.

d. a and b but not c.

e. a, b, and c.

Topic Test 1: Answers

1. **False.** It is true that behavior is a complex trait, which is influenced by the expression of large numbers of genes that play various roles in neural circuits. However, many instances exist in which mutations of single genes have large effects on behavior.

2. **True.** *shaker* is a strain of fly in which expression of a voltage-dependent potassium channel is disrupted. *shibire* is a *Drosophila* strain in which flies become paralyzed because of disrupted recycling of synaptic vesicles.

3. **False.** In fact, the amino acid sequence for a protein is now most often deduced from the nucleotide sequence of the gene encoding the protein, not the other way around.

4. **True.** Transgenic is a term that refers to the transfer of novel DNA into the genome of an organism. Originally, the novel DNA was commonly a gene from another species, so that the engineered transgenic organism would produce an alien protein. Also, in its early forms, the DNA incorporated randomly into the host genome, producing unpredictable results. More recently, transgenic organisms are produced using a technique called homologous recombination, in which the DNA incorporated into the host closely resembles a native piece of DNA. The introduced DNA then inserts into the host DNA only at the site of the native gene.

5. **True.** A "knockout" refers to the introduction of an altered gene to replace the native gene of an organism in such a way that the production of a functional protein from the native gene is prevented. Production of a knockout involves homologous recombination, in which the inserted DNA replaces the DNA around the coding region of the normal gene. The inserted DNA is similar to the native DNA, but the coding portion (exons) of the gene are truncated or eliminated entirely. A potential problem is that the disrupted gene is absent throughout embryonic development, and so functional deficits can result from indirect actions rather than from a direct role of the missing protein in the behavior. Also, other genes can compensate for the missing protein, masking the role of the missing protein. To circumvent these problems, researchers have developed more sophisticated knockout techniques, in which disruption of the gene is restricted to particular target cells (targeted knockouts) or in which the disruption of gene is triggered in the adult by an experimentally applied stimulus (inducible knockouts).

6. **a.** Because of its small size, short life cycle, and prolific reproduction, *D. melanogaster* has long been a standard for classical genetics, and in fact many of the principles for genetic analysis of inherited traits were first worked out for this species. Therefore, experimental tools based on this genetic analysis allow a number of advantageous shortcuts for the identification of mutated genes in the fruitfly. Statement b is incorrect because the *Drosophila* genome is actually substantially simpler than that of mammals. Statement c is incorrect because technologies for production of transgenic animals are actually more advanced for mice than for *Drosophila* (although the ease of genetic manipulations makes the need for transgenic capability less pressing in *Drosophila*).

TOPIC 2: FUNCTIONAL EXPRESSION OF GENES AND MUTAGENESIS

KEY POINTS

✓ *What is heterologous expression of a protein?*

✓ *How can a cell be induced to make a protein not normally found in the cell?*

✓ *What is site-directed mutagenesis?*

Expression of a protein where it is not normally found is called **heterologous expression**. To express a protein, the **messenger RNA (mRNA)** for the protein can be injected into a cell, whose protein translation machinery then synthesizes the protein. Frog oocytes (immature eggs) are commonly used for this purpose, because the cells are large and have abundant protein synthesis machinery. Another method for heterologous expression of a protein is to incorporate DNA that codes for the protein into cultured cells, with the transcription of the DNA into mRNA driven by a viral promoter merged to the DNA. Incorporating alien DNA into host cells is called **transfection**, and the transfection may be either temporary (transient transfection) or permanent (stable transfection). In some cases, specially engineered viruses have been used to introduce DNA into cells, taking advantage of the infection mechanisms that allow viruses to invade host cells.

A common use of heterologous expression is to link a particular neuronal characteristic to a particular gene expressed in the neuron. However, introduction of engineered DNA also offers the possibility of establishing which part of a protein molecule carries out a particular function of the protein. The DNA coding for the protein can be altered experimentally to change the amino acid sequence before the DNA is transfected into cells or transcribed into mRNA to be injected into oocytes. In this way, selected mutations can be introduced into the resulting protein to see how the function of the protein is altered. This procedure is called **site-directed mutagenesis**, which has been used extensively to determine the amino acids important in the functional characteristics of ion channels. For example, site-directed mutagenesis, followed by heterologous expression in oocytes, was used to establish that positively charged amino acids in a particular part of the molecule are responsible for the voltage-sensitivity of voltage-gated sodium channels of the action potential. The amino acid residues that determine the ionic selectivity of ion channels have also been studied using site-directed mutagenesis.

Topic Test 2: Functional Expression of Genes and Mutagenesis

True/False

1. A protein expressed only in the nervous system and not in other tissues may not be able to function properly if it is transfected into nonneuronal cells.

2. Heterologous expression of a protein requires that the protein must be synthesized artificially and then the protein is injected into the host cell.

3. RNA injected into a cell can be used to increase production of a protein but not to inhibit production of a protein.

4. Heterologous expression of a protein can be induced by injecting either RNA or DNA into the target cell.

5. Site-directed mutagenesis refers to random mutations introduced into transfected DNA or RNA by the target cell, after the nucleic acid has been injected into the host cell.

6. When the amino acid sequence of a protein is altered by site-directed mutagenesis, the resulting protein is always not functional.

Multiple Choice

7. In heterologous expression,
   a. a host cell should be selected that does not normally express the protein of interest.
   b. the host cell should be chosen to allow the functional characteristics of the protein to be more easily assessed.
   c. the host cell is always some type of immature egg cell, such as a frog oocyte.
   d. a and b but not c.
   e. a, b, and c.

Topic Test 2: Answers

1. **True.** The functional properties of an individual protein in a cell often depend on interactions with other proteins or on appropriate modifications made to the protein after it is produced by the ribosomes (posttranslational modifications). Although heterologous expression often produces proteins that work properly in their new environment, in some cases heterologous expression of a protein in cells that do not normally produce it fails to produce a functional protein or produces a protein whose functional characteristics are abnormal.

2. **False.** To induce a cell to manufacture a protein that it does not normally possess, researchers typically synthesize either DNA or mRNA, which is then injected into the target cell. In this way, the machinery for protein synthesis in the cell is used to generate the desired protein.

3. **False.** Like DNA, RNA consists of a sequence of nucleotide bases that can interact in pairs with complementary bases (A-T and C-G for DNA, A-U and C-G for RNA). Normally, only one strand of the double-stranded DNA in the nucleus is used as a template to synthesize mRNA during gene transcription. The resulting mRNA (called the sense strand) is complementary to the transcribed DNA strand and is used to guide protein translation. However, if a complementary strand of RNA to the sense strand (an antisense strand) is artificially introduced into the cell, it binds strongly to the sense strand and prevents the use of the sense mRNA for protein translation. Therefore, it is possible to inhibit the production of a protein by introducing antisense nucleotide sequence.

4. **True.** Injection of both RNA and DNA are commonly used to induce heterologous expression of proteins. In the case of RNA, mRNA is synthesized chemically from a DNA template for the protein of interest, and the mRNA is then injected into a cell that uses the mRNA to direct the synthesis of the protein, such as a frog oocyte. In the case of DNA, the DNA template for the protein is synthesized and then linked to a viral DNA promoter sequence. The promoter/DNA construction is then incorporated into the target

cell, where the promoter directs the transcription machinery of the cell to produce mRNA from the DNA template.

5. **False.** Mutations are not usually produced in the transfected DNA or RNA after it has been introduced into the target cells. Instead, site-directed mutagenesis is an experimental technique that allows experimenters to make well-defined specific changes in the nucleotide sequence to alter the amino acid sequence of the protein produced by the cell from the injected message.

6. **False.** Although site-directed mutagenesis may result in a nonfunctional protein, this is usually not the goal of the technique. Instead, researchers make changes in amino acids within a protein to obtain information about the function of particular regions of the protein.

7. **d.** The experimental goal of heterologous expression is to learn about the physiological function of a protein by placing it in a situation where it is not normally found. This usually requires that the host cell should not express the protein, so that experimenters can discern the alteration in physiological properties of the host induced by the expression of the protein. Similarly, the host cells should allow experimenters to more readily measure or analyze the relevant biochemical or physiological parameters expected to be affected by the test protein. Therefore, statements a and b are both correct. Statement c is false because many types of cells are used for heterologous expression studies.

IN THE CLINIC

In the study of human genetic diseases, a technique called positional cloning is used to identify the genetic defect underlying inherited ailments. The starting point for identification of the gene is collection of DNA from families afflicted by the disease. The DNA is then analyzed for different genetic markers, which are molecular signposts located at different positions along the 23 human chromosomes, to determine which markers are found among patients who have the disease but not in relatives who do not. A marker that is tightly associated with the disease focuses the search for the mutated gene to a particular location along a specific chromosome. After the general location is identified, a more detailed map of markers is made for that region, and markers linked to the disease are again identified. After the disease gene is localized to a sufficiently small stretch of the genome (about a million base pairs out of approximately 3 billion base pairs in the human genome), candidate genes can be identified either by sequencing the DNA base by base to determine the genes present in the region of interest or by making use of the sequence data already present in the genome database prepared by the Human Genome Project. The sequence of a candidate gene can then be directly compared in family members with and without the disease to establish the causal mutation.

Chapter Test

True/False

1. A spontaneous inherited mutation of a single gene can be used to localize the gene on a particular chromosome.

2. *timeless* and *period* are *Drosophila* mutants that have defective circadian rhythms.

3. *reeler* is a *Drosophila* mutant that has defective locomotion.

4. A transgenic animal is one in which half of the chromosomes come from one species and the other half from another species.

5. In a knockout mouse, the normal gene for a protein is replaced by a nonfunctional DNA sequence.

6. In a transgenic animal, the change in the genome may either add a novel gene or remove an existing gene.

7. Genetic analysis of molecules important for the nervous system can be carried out only in *Drosophila*.

8. To inhibit the production of a protein using antisense nucleotide sequence, it is necessary to know the complete gene sequence for the protein.

9. To express a protein in a cell that does not normally express it, it is necessary to know the complete gene sequence for the protein.

10. Heterologous expression refers to the expression of a protein in a cell that does not normally produce that protein.

11. In site-directed mutagenesis, the sequence of nucleotides is altered in the DNA encoding a protein to change the codon at a particular location.

12. Injection of mRNA into an oocyte will produce a transgenic animal if the egg is fertilized and allowed to grow into an adult animal.

Multiple Choice

13. The physiological function of a protein whose gene has been identified could be established by
    a. constructing a transgenic mouse in which the gene is knocked out.
    b. expressing the protein in frog oocytes.
    c. determining which chromosome contains the gene.
    d. a and b but not c.
    e. a, b, and c.

14. Positional cloning involves
    a. transgenic animals.
    b. establishing the location of a gene on a particular chromosome.
    c. site-directed mutagenesis.
    d. all of the above.
    e. none of the above.

15. Production of a transgenic animal involves
    a. homologous recombination.
    b. heterologous expression.
    c. injection of mRNA into oocytes.
    d. injection of expressed protein into oocytes.
    e. all of the above.

Short Answer

16. Name two *Drosophila* mutations that affect the nervous system and describe the functional role of the protein encoded by the mutated gene in each case.

Chapter Test Answers

1. **T** 2. **T** 3. **F** 4. **F** 5. **T** 6. **T** 7. **F** 8. **F** 9. **T** 10. **T** 11. **T** 12. **F**

13. **d** 14. **b** 15. **a**

16. Shaker: a mutation in a voltage-dependent potassium channel, which is involved in action potential repolarization in neurons and muscle cells. Shibire: a mutation in a GTPase required for endocytosis of vesicle membrane after neurotransmitter and therefore for vesicle recycling.

Check Your Performance:

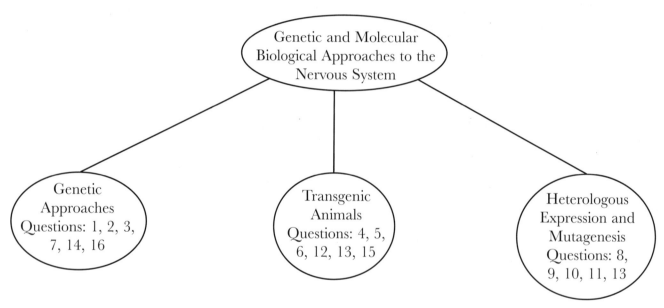

Note the number of questions in each grouping that you got wrong on the chapter test. Identify areas where you need further review and go back to relevant parts of this chapter.

Further help: If you continue to have difficulty with the concepts, a detailed presentation of these topics can be found in chapters 4 and 63 of *Principles of Neural Science, Third Edition*, by Kandel et al.

Second Midterm Exam

True/False

1. Contraction of a skeletal muscle fiber occurs when thick and thin filaments of a myofibril slide past each other, which is triggered when calcium ions are released from the sarcoplasmic reticulum and bind to the regulatory protein, troponin, associated with the thin filaments. (Chapter 11)

2. As tension in a muscle increases, the first motor units to be recruited at low levels of tension are the largest motor units, which have the largest number of muscle cells contacted by a single motor neuron. (Chapter 11)

3. In the myotatic reflex, the sensory neurons from the muscle spindles make direct inhibitory synapses onto the motor neurons of antagonistic muscles. (Chapter 12)

4. If the connection between the brain and spinal cord is severed, the spinal cord still retains the ability to produce locomotor movements. (Chapter 12)

5. The primary motor cortex, located in the precentral gyrus of the cerebral cortex, has a somatotopic organization, with the lower limb represented near the midline and the head represented at the lateral edge. (Chapter 13)

6. The three nuclei making up the basal ganglia are the caudate, the globus pallidus, and the substantia nigra. (Chapter 13)

7. The cell bodies of autonomic motor neurons are located in autonomic ganglia distributed throughout the body. (Chapter 14)

8. In the cardiac baroreceptor reflex, preganglionic neurons found in the intermedio-lateral cell column of the spinal cord are inhibited in response to increased blood pressure. (Chapter 14)

9. All motor neurons that control the eye muscles are located in the oculomotor nucleus (the nucleus of cranial nerve III) in the brainstem. (Chapter 15)

10. In the saccade control system of the brain, omnidirectional pause neurons stop firing during saccades of all directions, and the direction of a saccade is determined by which set or sets of excitatory burst neurons are activated by eye movement command centers. (Chapter 15)

11. The limbic system refers to the part of the neocortex concerned with emotion. (Chapter 16)

12. All neurosecretory neurons of the hypothalamus release the substances they secrete directly into blood vessels from their nerve endings. (Chapter 16)

13. In almost all people, loss of the ability to speak would be caused by damage to the left cerebral hemisphere. (Chapter 17)

14. Agnosia refers to the loss of ability to identify objects based on particular sensory information, without damage to the corresponding primary sensory cortex. (Chapter 17)

15. If the hippocampus is destroyed, short-term memory and existing long-term memory are relatively intact, but the ability to transfer new information into long-term memory is impaired. (Chapter 18)

16. Blocking protein synthesis would prevent long-term sensitization but would have no effect on short-term sensitization in the *Aplysia* withdrawal reflex. (Chapter 18)

17. The extracellular matrix in all parts of the body has the same molecular composition, and all neurons in the nervous system express the same types of adhesion molecules that interact with the extracellular matrix. (Chapter 19)

18. Immature muscle cells insert acetylcholine molecules at high density at a single site that will become the neuromuscular junction, and growth cones of motor neurons are then attracted to this site during embryonic development. (Chapter 19)

19. A transgenic organism is one in which non-native DNA has been incorporated into the genome of the organism. (Chapter 20)

20. Heterologous expression refers to the artificial synthesis of a protein, which is then injected into a cell that does not normally express that protein. (Chapter 20)

Multiple Choice

21. The nervous system controls the strength of contraction of a skeletal muscle by (Chapter 11)
    a. varying the total number of motor neurons activated, thus changing the total number of motor units contracting.
    b. varying the frequency of action potentials in the motor neuron of a single motor unit.
    c. varying the number of muscle cells that contract within a single motor unit.
    d. a and b, but not c.
    e. a, b, and c.

22. The functional roles of the myotatic and inverse myotatic reflexes are (Chapter 12)
    a. The myotatic reflex maintains constant muscle tension, and the inverse myotatic reflex maintains constant muscle length.
    b. The myotatic reflex controls the length of extensor muscles, and the inverse myotatic reflex controls the length of flexor muscles.
    c. The myotatic reflex stimulates muscles that withdraw the ipsilateral limb and extend the contralateral limb, and the inverse myotatic reflex stimulates muscles that withdraw the contralateral limb and extend the ipsilateral limb.
    d. The myotatic reflex maintains constant muscle length, and the inverse myotatic reflex maintains constant muscle tension.
    e. None of the above.

23. The corticospinal tract (Chapter 13)
    a. is involved in gross motor control such as locomotion, rather than fine motor control such as manipulation of objects with the fingers.
    b. consists of axons that project from the motor cortex to the brainstem, where they synapse on interneurons that relay the motor commands to the spinal cord.
    c. originates from pyramidal neurons found in layer V of the motor cortex.
    d. All of the above.
    e. None of the above.

24. Acetylcholine is the neurotransmitter used by (Chapter 14)
    a. somatic motor neurons

b. parasympathetic motor neurons

c. parasympathetic preganglionic neurons

d. sympathetic preganglionic neurons

e. All of the above

25. In the vestibulo-ocular reflex (Chapter 15)

 a. sensory neurons in the vestibular ganglion of the semicircular canals make direct synaptic connections with motor neurons of the ocular muscles.

 b. motor neurons of the eye muscles are located in three different nuclei in the brainstem: the oculomotor nucleus, the trochlear nucleus, and the abducens nucleus.

 c. neurons of the vestibular nuclei in the brainstem receive sensory inputs only from the vestibular apparatus.

 d. horizontal motion of the head to the left causes the eyes to move to the left in the horizontal plane.

 e. the amount of movement of the eye is controlled by feedback from muscle stretch receptors of the ocular muscles to ensure that rotation of the eyes matches rotation of the head.

26. In the hypothalamus (Chapter 16)

 a. both the lateral hypothalamic area and the ventromedial nucleus are involved in feeding.

 b. hormones of the anterior pituitary are secreted directly into the bloodstream by hypothalamic neurosecretory neurons.

 c. the anterior nuclei of the hypothalamus are specialized for control of body temperature, and the posterior nuclei of the hypothalamus are specialized for control of water intake.

 d. All of the above.

 e. None of the above.

27. In aphasia (Chapter 17)

 a. damage to Wernicke's area would most likely be associated with expressive aphasia.

 b. damage to Broca's area would most likely be associated with receptive aphasia.

 c. damage to the left temporal lobe would likely produce receptive aphasia in a right handed person.

 d. All of the above.

 e. None of the above.

28. Long-term potentiation in the hippocampus requires (Chapter 18)

 a. postsynaptic depolarization

 b. presynaptic action potential activity

 c. influx of calcium into the postsynaptic cell

 d. All of the above

 e. None of the above

29. The extracellular matrix can include the following molecular components (Chapter 19)

 a. collagen

 b. glycosaminoglycans

 c. glycoproteins

 d. laminin

 e. All of the above.

30. The functional role of a protein in the nervous system could be established by (Chapter 20)
    a. constructing a transgenic animal in which the gene encoding the protein is removed.
    b. identifying the gene for the protein and determining the DNA sequence of the gene.
    c. determining which neurons in the nervous system express mRNA encoding the protein.
    d. identifying the chromosome carrying the gene encoding the protein.
    e. All of the above.

Answers

1. **T**   2. **F**   3. **F**   4. **T**   5. **T**   6. **F**   7. **T**   8. **T**   9. **F**   10. **T**   11. **F**   12. **T**   13. **T**   14. **T**   15. **T**   16. **T**   17. **F**   18. **F**   19. **T**   20. **F**   21. **d**   22. **d**   23. **c**   24. **e**   25. **b**   26. **a**   27. **c**   28. **d**   29. **e**   30. **a**

Final Exam

True/False

1. The cerebrospinal fluid fills the hollow core of the central nervous system: the cerebral ventricles in the brain and the spinal canal in the spinal cord. (Chapter 1)

2. The notochord is the structure that gives rise to the spinal cord during embryonic development. (Chapter 2)

3. The Nernst equation gives the equilibrium potential for an ion: the value of membrane potential at which the electrical gradient across the membrane exactly compensates for the concentration gradient for the ion. (Chapter 3)

4. The depolarizing phase of the action potential is caused by the explosive cycle of depolarization opening voltage-dependent sodium channels, which in turn produces further depolarization. (Chapter 4)

5. An inhibitory postsynaptic potential would result from a neurotransmitter that opens nonspecific cation channels that are equally permeable to sodium and potassium ions. (Chapter 5)

6. The receptive field of a sensory neuron refers to the portion of the cerebral cortex that receives inputs from the neuron. (Chapter 6)

7. The dorsal columns of the spinal cord consist predominantly of ascending axons of spinal neurons that receive incoming sensory information about pain and temperature. (Chapter 7)

8. Many neurons in the primary visual cortex respond best to bars or stripes of light of a particular orientation. (Chapter 8)

9. In the cochlea, hair cells at the basal end respond best to low-frequency sound, whereas hair cells at the apical end respond best to high-frequency sound. (Chapter 9)

10. In the olfactory system, each olfactory receptor neuron has an axon that projects directly to a particular glomerulus in the olfactory bulb. (Chapter 10)

11. In skeletal muscle fibers, thick filaments are made up of the protein actin, and thin filaments are made up of the protein myosin. (Chapter 11)

12. In the inverse myotatic reflex, primary sensory neurons from a particular muscle make direct excitatory synapses onto motor neurons of the same muscle. (Chapter 12)

13. The basal ganglia are the caudate, the putamen, and the globus pallidus. (Chapter 13)

14. The sympathetic and somatic divisions are the two divisions of the autonomic nervous system. (Chapter 14)

15. Motor neurons that control the eye muscles are located in three cranial nerve nuclei: the oculomotor nucleus (cranial nerve III), the trochlear nucleus (cranial nerve IV), and the abducens nucleus (cranial nerve VI). (Chapter 15)

16. The Papez circuit is another name for the limbic system. (Chapter 16)

17. Broca's area is part of the frontal lobe, and damage to Broca's area causes difficulty in speech production. (Chapter 17)

18. Destruction of the hippocampus would erase all memories. (Chapter 18)

19. Cellular adhesion molecules interact with complementary adhesion molecules on the surface of surrounding cells or in the extracellular matrix. (Chapter 19)

20. Information about the function of a gene in the nervous system can be obtained if either a naturally occurring or induced mutation in the gene can be identified. (Chapter 20)

Multiple Choice

21. The forebrain includes the following structures (Chapter 1)
    a. The diencephalon
    b. The cerebral cortex
    c. The basal ganglia
    d. All of the above
    e. None of the above

22. During development of the nervous system (Chapter 2)
    a. the neurons of the central nervous system arise from the neural crest, and the neurons of the peripheral nervous system arise from the neural tube.
    b. the three vesicles that give rise to the brain are called the telencephalon, diencephalon, and the cerebrum.
    c. the neuroectoderm gives rise to the cells of the nervous system.
    d. All of the above
    e. None of the above

23. Suppose a neuron is permeable to sodium and potassium ions and that the intracellular and extracellular fluids have their usual compositions. If p_{Na} were doubled, leaving p_K unchanged, the membrane potential of the neuron would (Chapter 3)
    a. become more positive (depolarize).
    b. become more negative (hyperpolarize).
    c. remain unchanged.

24. The refractory period following an action potential arises because (Chapter 4)
    a. voltage-dependent potassium channels close slowly upon repolarization and so remain open for a period after the action potential.
    b. the inactivation gates (h-gates) of voltage-dependent sodium channels reopen slowly upon repolarization, and so the sodium channels are unable to open again upon depolarization for a period after the action potential.
    c. the activation gates (m-gates) of voltage-dependent sodium channels close slowly upon repolarization, and so the sodium channels remain open for a period after the action potential.
    d. both voltage-dependent sodium channels and voltage-dependent potassium channels are incapable of responding again to depolarization for a period after the action potential.
    e. None of the above.

25. Neurotransmitter release (Chapter 5)
    a. occurs when membrane-bound synaptic vesicles are released from the presynaptic cell and then fuse with the plasma membrane of the postsynaptic cell.
    b. requires the release of calcium ions from the postsynaptic cell.

c. is required for excitatory neurotransmission but not for inhibitory neurotransmission.

d. All of the above

e. None of the above

26. The receptor potential of a primary sensory neuron (Chapter 6)

 a. is another name for the action potential in a primary sensory neuron.

 b. is the change in membrane potential resulting from the process of sensory transduction.

 c. is stimulated by synaptic inputs from other primary sensory neurons.

 d. All of the above are correct.

 e. None of the above is correct.

27. The primary somatosensory cortex (Chapter 7)

 a. is located in the postcentral gyrus, just behind the central sulcus of the cerebral cortex.

 b. is organized in a somatotopic map, with sensory inputs from the head represented nearest the midline and inputs from the feet at the most lateral position.

 c. receives direct synaptic input from primary sensory neurons, without any intervening synapses.

 d. on the right side of the brain receives sensory information from the right side of the body.

 e. receives sensory information only from the dorsal column pathway of the spinal cord.

28. In the retina (Chapter 8)

 a. photoreceptors depolarize in response to illumination.

 b. light is absorbed in rod photoreceptors by the visual pigment molecule, rhodopsin.

 c. center-surround receptive fields are observed only in retinal ganglion cells.

 d. All of the above

 e. None of the above

29. The primary auditory cortex (Chapter 9)

 a. receives direct synaptic inputs from the cochlear nucleus, without intervening synapses.

 b. is located in the frontal lobe of the cerebral cortex.

 c. is organized in a two-dimensional array of vertically oriented frequency columns and binaural columns.

 d. All of the above

 e. None of the above

30. The olfactory bulb (Chapter 10)

 a. receives olfactory inputs from the thalamus, which in turn receives direct synaptic input from the axons of primary olfactory receptor neurons.

 b. receives synaptic inputs from the glomeruli of the olfactory paleocortex.

 c. is located within the nasal cavity and contains the olfactory epithelium.

 d. All of the above

 e. None of the above

31. During contraction of a skeletal muscle cell (Chapter 11)

 a. calcium ions are released from the transverse tubules to initiate contraction.

 b. calcium ions bind to the regulatory protein tropomyosin, which is associated with the thin filaments.

c. neither the thin filaments nor the thick filaments change in length.

d. All of the above

e. None of the above

32. The myotatic reflex differs from the inverse myotatic reflex in which of the following ways? (Chapter 12)

a. The myotatic reflex produces limb flexion, whereas the inverse myotatic reflex produces limb extension.

b. The sensory neurons of the myotatic reflex innervate muscle spindles, whereas the sensory neurons of the inverse myotatic reflex innervate Golgi tendon organs.

c. The myotatic reflex is activated by muscle stretch, whereas the inverse myotatic reflex is activated by muscle relaxation.

d. All of the above

e. None of the above

33. The primary motor cortex (Chapter 13)

a. has output neurons whose axons form the pyramidal tract.

b. has output neurons that make direct synaptic connections with motor neurons in the spinal cord.

c. is located in the precentral gyrus of the cerebral cortex.

d. All of the above

e. None of the above

34. Neurons of the intermediolateral cell column of the spinal cord make direct synaptic connection with (Chapter 14)

a. paravertebral sympathetic ganglia

b. prevertebral sympathetic ganglia

c. sympathetic chain ganglia

d. All of the above

e. None of the above.

35. During a saccade (Chapter 15)

a. neural circuits in the superior colliculus help determine the direction of the eye movement.

b. omnidirectional pause neurons fire a rapid burst of action potentials.

c. excitatory burst neurons stop firing.

d. inhibitory burst neurons stop firing.

e. All of the above

36. Which of the following hormones are released in the posterior lobe of the pituitary? (Chapter 16)

a. Growth hormone and thyroxin

b. Prolactin and somatotropin

c. Thyrotropin and ACTH

d. The gonadotropins LH and FSH

e. Vasopressin and oxytocin

37. Which of the following brain regions, lobes, and neurological syndromes are associated? (Chapter 17)

a. Wernicke's area; frontal lobe; expressive aphasia

b. Broca's area; temporal lobe; receptive aphasia

    c. Angular gyrus; parietal lobe; agnosia

    d. Wernicke's area; frontal lobe; agnosia

    e. Broca's area; parietal lobe; receptive aphasia

38. Long-term potentiation in a hippocampal neuron (Chapter 18)

    a. is thought to involve changes in both postsynaptic and presynaptic events at the potentiated synapse, resulting in enhanced synaptic transmission.

    b. occurs only at synapses that are silent during a period of strong activity in other synapses on the same postsynaptic neuron.

    c. produces an enhancement of the strength of synaptic transmission that lasts for a few hours.

    d. All of the above

    e. None of the above

39. The guidance of growing neurites during development of the nervous system involves (Chapter 19)

    a. interaction of adhesion molecules on the surface of the neurite with complementary adhesion molecules in the extracellular matrix.

    b. interaction of adhesion molecules on the surface of the neurite with complementary adhesion molecules on neighboring cells.

    c. chemotropic molecules that attract or repel growth cones.

    d. All of the above

    e. a and b, but not c

40. Regarding genetic and molecular biological analysis of the nervous system (Chapter 20)

    a. genetic analysis is possible in simple organisms such as *Drosophila*, but not in more complex organisms, such as mammals.

    b. in a transgenic animal, the inserted DNA inserts randomly in the genome of the host, and it is not possible to control the site of insertion.

    c. it is possible to test the function of a protein by selectively disrupting the corresponding gene by replacing it with a nonfunctional gene that does not encode a functional protein.

    d. All of the above

    e. a and b, but not c

Answers

1. **T** 2. **F** 3. **T** 4. **T** 5. **F** 6. **F** 7. **F** 8. **T** 9. **F** 10. **T** 11. **F** 12. **F**

13. **T** 14. **F** 15. **T** 16. **F** 17. **T** 18. **F** 19. **T** 20. **T** 21. **d** 22. **c** 23. **a**

24. **b** 25. **e** 26. **b** 27. **a** 28. **b** 29. **c** 30. **e** 31. **c** 32. **b** 33. **d** 34. **d**

35. **a** 36. **e** 37. **c** 38. **a** 39. **d** 40. **c**